Dynamical Problems of Rigid-Elastic Systems and Structures

International Union of Theoretical and Applied Mechanics

N. V. Banichuk · D. M. Klimov
W. Schiehlen (Eds.)

Dynamical Problems of Rigid-Elastic Systems and Structures

IUTAM Symposium, Moscow, USSR
May 23 – 27, 1990

Springer-Verlag
Berlin Heidelberg New York
London Paris Tokyo
Hong Kong Barcelona Budapest

Prof. N.V. Banichuk
Prof. D. M. Klimov

Institute for Problems in Mechanics
USSR Academy of Sciences
Pr. Vernadskogo 101
Moscow 117526
USSR

Prof. W. Schiehlen

Institute B of Mechanics
University of Stuttgart
Pfaffenwaldring 9
W-7000 Stuttgart 80
Germany

ISBN 3-540-53788-0 Springer-Verlag Berlin Heidelberg NewYork
ISBN 0-387-53788-0 Springer-Verlag NewYork Berlin Heidelberg

Typesetting: Camera ready by authors
Offsetprinting: Color-Druck Dorfi GmbH, Berlin; Bookbinding: Lüderitz & Bauer, Berlin
61/3020 543210 – Printed on acid-free paper

Preface

The International Union of Theoretical and Applied Mechanics (IUTAM) initiated and supported an International Symposium on Dynamical Problems for Rigid-elastic Systems and Structures held in 1990 in Moscow, USSR. The Symposium was intended to bring together scientists working in the fields of multibody system dynamics and finite element systems with special emphasis to modeling, simulation, optimization and control.

A Scientific Committee was appointed by the Bureau of IUTAM with following members:

N.V. Banichuk (USSR),
E.J. Haug (USA),
Y. Hori (Japan),
S. Kaliszky (Hungary),
D.M. Klimov (USSR), Chairman,
L. Lilov (Bulgaria),
F. Niordson (Denmark),
B. Roth (USA),
W. Schiehlen (Germany),
G. Schmidt (Germany),
J. Wittenburg (Germany).

The chairman invited the participants on recommendation by the Scientific Committee. As a result 48 active scientific participants from 11 countries followed the invitation, and 32 papers were presented in lecture sessions. The available manuscripts were reviewed by the Scientific Committee after the Symposium, and 24 of them are collected in this volume. At the Symposium a tour to the Institute for Problems of Mechanics, USSR Academy of Sciences, was arranged.

The scientific lectures were devoted to the following topics:

- Modeling and Optimization,
- Dynamics of Systems with Elastic Constraints,
- Vibrations,
- Multibody Systems.

The papers in this volume are arranged in alphabetical order with respect to the authors. They indicate the importance of multibody and finite element modeling with respect to many applications in mechanical engineering. The phenomena of elasticity or flexibility of the systems, respectively, require improved simulation and optimization methods. Further, the aspects of control engineering in highly dynamical systems are of major importance. The presentations and discussions during the Symposium will certainly stimulate future investigations in the field of mechanical systems. The publication of the proceedings will contribute to this development.

Generous financial and material support was of great help for carrying out the Symposium. The help of two sponsors is gratefully acknowledged:

International Union of Theoretical and Applied Mechanics (IUTAM),

Institute for Problems of Mechanics, USSR Academy of Sciences.

The Symposium was arranged by a Local Organizing Committee recruited from the Institute for Problems of Mechanics under the chairmanship of N.V. Banichuk and D.M. Klimov. The social programme included a banquet and an excursion providing excellent opportunities for further scientific discussions.

In the editorial work for this volume the help of D. Bestle from the Institute B of Mechanics, University of Stuttgart was especially valuable. In addition, we would like to express our sincere thanks to all members of the Local Organizing Committee who contributed to the Symposium. Furthermore, thanks are due to the Springer-Verlag for the sympathetic cooperation.

Moscow and Stuttgart, December 1990

N.V. Banichuk, D.M. Klimov, W. Schiehlen

List of Participants

(Chairmen are identified by an asterisk)

Akulenko, L.D., Institute for Problems in Mech., USSR Academy of Sciences, pr. Vernadskogo 101, Moscow 117526, USSR

* **Banichuk, N.V.**, Institute for Problems in Mech., USSR Academy of Sciences, pr. Vernadskogo 101, Moscow 117526, USSR

Beletsky, V.V., Keldysh-Institute of Appl. Math., USSR Academy of Sciences, Miusskaya pl. 4, Moscow 125047, USSR

Borri, M., Dipartimento di Ingegneria Aerospaziale, Politecnico di Milano, Milano 20133, Italy

Bratus, A.S., Dept. of Appl. Math., Moscow Railway Eng. Institute, Moscow, USSR

Bremer, H., Lehrstuhl B für Mechanik, Technische Universität München, Arcisstr. 21, W-8000 München 2, Germany

* **Chernousko, F.L.**, Institute for Problems in Mech., USSR Academy of Sciences, pr. Vernadskogo 101, Moscow 117526, USSR

Churkina, N.I., Moscow Aviation Institute, Moscow, USSR

* **Crandall, S.H.**, Dept. of Mech. Engineering, Massachusetts Institute of Technology, Cambridge, MA 02139, USA

Dehombreux, P., Service de Mécanique Rationnelle, Faculté Polytechnique de Mons, Boulevard Dolez 31, 7000 Mons, Belgium

Demin, V.G., Moscow State University, Moscow, USSR

Frolov, K.V., Mechanical Engineering Research Institute, USSR Academy of Sciences, 4 Griboedova, Moscow 101830, USSR

Furta, S.D., Moscow Aviation Institute, Moscow, USSR

Goureyev, L.V., Moscow Aviation Institute, Moscow, USSR

Ishlinsky, A.Yu., Institute for Problems in Mech., USSR Academy of Sciences, pr. Vernadskogo 101, Moscow 117526, USSR

Karapetyan, A.V., Computing Centre of the USSR Academy of Sciences, 40 Vavilov Street, Moscow 117967, USSR

* **Klimov, D.M.**, Institute for Problems in Mech., USSR Academy of Sciences, pr. Vernadskogo 101, Moscow 117526, USSR

Kozlov, R.I., Irkutsk Center for Computations, Sib. Dept. of USSR Academy of Sciences, Irkutsk, USSR

Kravchenko, S.V., Mechanical Engineering Research Institute, USSR Academy of Sciences, 4 Griboedova, Moscow 101830, USSR

Kreuzer, E.J., Meerestechnik II - Strukturmechanik, TU Hamburg-Harburg, Eißendorfer Straße 42, W-2100 Hamburg 90, Germany

Levin, E.M., Keldysh-Institute of Appl. Math., USSR Academy of Sciences, Miusskaya pl. 4, Moscow 125047, USSR

Lilov, L., Institute of Math. and Mech. with Comp. Center, Bulgarian Academy of Sciences, ul. 36, bl. 8, P.O.B. 373, 1000 Sofia, Bulgaria

Lin, Y.K., Center for Applied Stochastics Research, Florida Atlantic University, Boca Raton, FL 33431, USA

Luschin, L.P., Institute for Problems in Mech., USSR Academy of Sciences, pr. Vernadskogo 101, Moscow 117526, USSR

Magi, M., Chalmers University of Technology, S-41296 Goteborg, Sweden

Maißer, P., Akademie der Wissenschaften, Institut für Mechanik, Postfach 408, O-9010 Chemnitz, Germany

Markeev, A.P., Institute for Problems in Mech., USSR Academy of Sciences, pr. Vernadskogo 101, Moscow 117526, USSR

Markov, Yu.G., Moscow Aviation Institute, Moscow, USSR

Mikhajlov, F.A., Moscow Aviation Institute, Moscow, USSR

Mosolev, V.E., Moscow Aviation Institute, Moscow, USSR

Müller-Slany, H.H., Institut für Mechanik und Regelungstechnik, Universität-Gesamthochschule Siegen, W-5900 Siegen, Germany

Muravej, V.A., Moscow Institute of Aircraft Technology, Moscow, USSR

Nabullin, M.K., Irkutsk Center for Computations, Sib. Dept. of USSR Academy of Sciences, Irkutsk, USSR

Nash, W.A., College of Engineering, University of Massachusetts, Amherst, MA 01003, USA

* **Nesterov, S.V.**, Institute for Problems in Mech., USSR Academy of Sciences, pr. Vernadskogo 101, Moscow 117526, USSR

* **Pascal, M.**, Laboratoire de Modélisation en Mécanique, Universite Pierre et Marie Curie, Tour 66-4, place Jussieu, 75252 Paris Cédex 05, France

Pater, A.D. de, Delft University of Technology, Mekelweg 2, 2628 CD Delft, The Netherlands

Ryaboy, V.M., Mechanical Engineering Research Institute, USSR Academy of Sciences, 4 Griboedova, Moscow 101830, USSR

Sadov, Yu.A., Keldysh-Institute of Appl. Math., USSR Academy of Sciences, Miusskaya pl. 4, Moscow 125047, USSR

* **Santini, P.**, Scuola d'Ingeneria Aerospaziale, Dell Universita di Roma, 00184 Roma, Via Eudossiana 16, Italy

* **Schiehlen, W.**, Institut B für Mechanik, Universität Stuttgart, Pfaffenwaldring 9, W-7000 Stuttgart 80, Germany

Sharanjuk, A.V., Institute for Problems in Mech., USSR Academy of Sciences, pr. Vernadskogo 101, Moscow 117526, USSR

Somov, E.I., Irkutsk Center for Computations, Sib. Dept. of USSR Academy of Sciences, Irkutsk, USSR

Stejskal, V., Faculty of Mechanical Engineering, Czech Technical University of Prague, Technicka 4, 16607 Praha 6, CSFR

Stifter, S., Research Institute for Symbolic Computation, Johannes Kepler Universität, Schloß Hagenberg, A-4232 Hagenberg, Austria

Suetov, A.P., Institute of Mechanics and Mathematics, Ur. Dept. of USSR Academy of Sciences, Sverdlovsk, USSR

Valášek, M., Faculty of Mechanical Engineering, Czech Technical University of Prague, Technicka 4, 16607 Praha 6, CSFR

* **Wittenburg, J.**, Institut für Technische Mechanik, Universität Karlsruhe, Kaiserstr. 12, W-7500 Karlsruhe, Germany

Opening Address

D.M. Klimov, Symposium Chairman

Ladies and gentlemen!

Dear colleagues!

Today we open IUTAM Symposium on dynamical problems of rigid-elastic systems and structures. There had been a few symposiums in this field already. The first of these, under title "Dynamics of Multibody Systems", took place in Munich/Germany in 1977. The chairman of this conference was professor K.Magnus.

In response to industrial needs and as a result of the ever growing power of computers, the subject of mechanics of elastic-bodies-systems and structures is developing rapidly. Major steps are being taken to exploit new capabilities for improved modeling, computer-aided analysis and design, computerized formulation of equations of motion, and advanced solution techniques. Methods of structural optimization, control and identification are receiving extensive attention. Notable progress is achieved in dynamic studies of mechanical systems taking into account the effects of elasticity, viscosity, temperature deformations. Methods for dealing with deformable bodies and efficient determination of interaction forces are being developed particular in the fields of robots and large space structures.

Further development of these investigations is stipulated by the important problems of quality growing for mechanical systems, improving of dynamical characteristics, reduction of the cost, weight and material consumption for mechanical systems and structures.

The aims of the Symposium are to generate knowledge, to stimulate research, to disseminate new ideas, and to acquaint the scientific community in general with the work currently in progress in the area of dynamic of mechanical systems and structures. For the above reason, the following topics are proposed:

- Modeling in dynamics and the problems of identification for mechanical systems and structures. Effects of viscosity, elasticity and temperature strains.

- New methods and algorithms for computer-aided analysis of complicated elastic systems. Application of symbolic calculation.

- Design and optimization of mechanical systems and structures for dynamic condition.

I'd like to say that local committee tried to do all necessary preparations. And my colleagues from the Institute of Problems in Mechanics and I hope that this Symposium will be useful for the international scientific community.

So welcome everybody.

Opening Address

W. Schiehlen, Secretary–General of IUTAM

Mr. President, Mr. Chairman, Academicians,
Ladies and Gentlemen,

On behalf of the International Union of Theoretical and Applied Mechanics I would like to welcome you at the Symposium on

Dynamical Problems of Rigid-Elastic Systems and Structures.

The proposal to arrange this Symposium was accepted in August 1988 by the General Assembly of IUTAM at the Grenoble meeting. Since that time the chairman, Professor D.M. Klimov, and his colleague, Professor N.V. Banichuk, have done a hard job preparing a challenging scientific programme.

The International Union of Theoretical and Applied Mechanics expects that the members of the Scientific Committee participate in the Symposium, too. And therefore, I will especially mention the scientists: Professor Lilov from Bulgaria and Professor Wittenburg from West Germany (FRG).

All of you know that it is not easy to organize an international symposium. There have been problems in contacting the members of the Scientific Committee for selection of participants and papers. But finally looking at the result, a challenging scientific programme was obtained.

However, the real success of a Symposium can be achieved only if all participants use the opportunity to exchange ideas on new developments, this week especially with respect to dynamical problems of multibody systems, vibrations and optimization.

Therefore, I would like to encourage all participants to lively discussions. The USSR is a member of the International Union for Theoretical and Applied Mechanics since 1956 and it is one of the leading countries in mechanical sciences.

First, I have to mention the 13th World Congress on Mechanics held 1972 in Moscow. This congress was, even until today, the world's largest meeting in the field of mechanics.

Secondly, there have been organized 10 IUTAM Symposia in the USSR in Solid and Fluid Mechanics. And today, we call to order the eleventh IUTAM Symposium in this country. Especially, the activities of the Sowjet scientists in the field of multibody systems are honoured internationally by this Symposium.

The science of multibody systems dates back to the sixties of this century when the space programme required improved modeling, simulation and optimization technques. In connection with the increasing power of computational facilities during the last two decades, the multibody formalisms represent new tools for computer aided engineering, or CAE-techniques, respectively.

Two IUTAM Symposia have been devoted to multibody system dynamics. Professor Magnus organized 1977 a very successful symposium on this topic in Munich, FRG. Later in 1985 Professor Bianchi and myself held a IUTAM Symposium on multibody systems in Udine, Italy. And this week we will deal with recent developments in multibody systems combining rigid and elastic bodies as well as structures.

Ladies and Gentlemen, enjoy the honour and pleasure of being invited to a IUTAM symposium which means a first class symposium. Thank you.

Opening Address

K.V. Frolov, Deputy President of USSR Academy of Sciences

Dear Colleagues, Comrades,

I am glad to use this opportunity to greet you on the occasion of opening of the IUTAM Symposium on Dynamical Problems of Rigid-Elastic Systems and Structures.

The scope of this Symposium includes complex problems of mechanics. Their solutions provide scientific basis for further progress in most advanced and promising fields of modern technology, such as robotics, development of large space systems, in particular orbital structures, vibration technology and vibration protection systems.

Research and solution of dynamical problems in the above-mentioned fields demand the efforts of numerous and mighty teams of scientists. Research institutes of the USSR Academy of Sciences also pay much attention to the solution of these problems. The results of fundamental research of non-linear vibration made it possible to develop new promising wave technologies, and to create high-performance resonance robots and high-precision manipulators. Essential results have been also achieved in the theory of automatic control. The methods of synthesis of continuous non-linear control laws based on Lyapunov vector functions technique have been developed, as well as the method of stochastic programmed synthesis for the control with instantaneous and integral constraints.

Results of fundamental and applied research in the field of mechanics have been used in design and construction of space system "Energia - Buran", new aircraft IL-96-300, TU-204, AN-255 as well as in realization of FOBOS-2 spacecraft.

Nowadays, modern computers present a powerful additional means for theoretical investigations, simulations, calculations, optimal design and control. Their use enriches modern mechanics with new qualities and potential which were beyond the boldest dreams of scientists in the past.

Today one of the priority issues is the problem of safety in the wide sense of the word. This includes environmental safety, reliability of nuclear reactors and major power and transport systems, security of chemical industries, search and rescue of ships in trouble, faultless flights in civil aviation, ergonomic problems solution and more productive work of a human-operator.

In view of the large scope of research in dynamics devoted to the fundamental studies and solution of practical problems (especially those connected with safety insuring) summing up of results, analysis of recent achievements, and personal contacts with the outstanding world-wide known scientists acquire a special importance. The IUTAM Symposium provides favorable conditions for such an exchange of scientific ideas and results.

At this Symposium we hope to discuss, report and learn many latest results obtained in various fields of dynamics. The Symposium opens up opportunities for joint efforts and agreed activity of scientists belonging to different schools. In the reports and discussions at the Symposium original results and state-of-the-art reviews will be presented in a number of important branches of dynamics which attract now the attention of scientists from various countries.

Collaboration of scientists promotes both the progress of science and mutual understanding among nations in their strive for peace and security on Earth. Progress in international scientific and technical cooperation in future implies the creation of favorable conditions for the establishment of closer relations between Soviet enterprises and institutions, and their counterparts in Western Europe on the basis of common European home concept. It will also facilitate the establishment of joint ventures and common scientific programs involving the efforts of major companies, universities and societies in USA, Japan and other countries in the spirit of new thinking and priority of human values.

I wish you every success in your fruitful work at the Symposium for the benefit of science and mankind.

Contents

Oscillations of a Rigid Body with a Cavity Containing a Heterogeneous Fluid

L. D. AKULENKO, S.V. NESTEROV

The Institute for Problems in Mechanics
USSR Academy of Science, Moscow, USSR

Summary

Oscillations of the mechanical system with infinite (countable) number of degrees of freedom and internal resonance is considered . This system consists of a rigid body with a rectangular cavity containing a heavy heterogeneous ideal fluid. The body is attached to a fixed base by means of a spring and can move in a horizontal direction. Using asymptotic technique analytical solution for the equations describing joint oscillations of the body and the fluid was obtained in both resonant and nonresonant cases. Some qualitative features of the system were found. We present here not only theoretical results but also the experimental data which reveal beating and changing the frequency of oscillations of the body.

Analysis of oscillations

A heterogeneous fluid consisting of two layers of different densities $\rho_{1,2}$, $0 \le \rho_1 < \rho_2$ occupies a rectangular cavity inside the rigid body. Particles of the fluid are exerted by the gravity force. Depths of the layers are equal to h_1 and h_2; d and r are the length and the width of the cavity respectively. The fluid is assumed to be ideal and incompressible. The rigid body is exerted by an elastic force and can move in a horizontal direction. At an initial instant the body is given a small displacement s_0 from the equilibrium position, the boundary between the layers of fluids is horizontal and the whole system is at rest.

N.V. Banichuk · D. M. Klimov · W. Schiehlen (Eds.)
Dynamical Problems of Rigid-Elastic
Systems and Structures
IUTAM Symposium Moscow, USSR, May 23-27, 1990

To derive the equation of motion of the system "a rigid body + heterogeneous liquid" we firstly solve the problem of hydrodynamics for the internal motion of a heterogeneous fluid, provided the motion of a rigid body is specified. Then we calculate forces of the fluid. As a result we obtain the integrodifferential equation governing the motion of a rigid body with a cavity containing a heterogeneous fluid

$$\ddot{s} + \Omega^2 s = - \varepsilon \int_0^t R(t - \tau)\, \ddot{s}(\tau)\, d\tau , \tag{1}$$

$$R(t) = \frac{1}{\Omega^2} \sum_{k=0}^{\infty} \frac{\omega_{2k+1}^3}{(2k + 1)^4} \sin \omega_{2k+1} t .$$

Here s is the displacement of the rigid body from the equilibrium position;

R is the kernel of the integral operator, describing the influence of internal waves;

ω_{2k+1} are the eigenfrequencies of internal waves appearing on the intermediate surface;

$\Omega^2 = \lambda^2/M$ - the frequency of oscillations without influence of the waves;

λ^2 - the stiffness of an elastic link;

M is the efficient mass of the system "the rigid body + the fluid";

$\varepsilon = \dfrac{8rd^3}{\pi} \dfrac{(\rho_2 - \rho_1)\Omega^2}{Mg}$ $(\varepsilon > 0)$ - the numerical parameter, usually ε is small.

One has to solve the equation (1) with initial conditions

$$s(0) = s_0, \qquad \dot{s}(0) = 0 . \tag{2}$$

Formally the solution of the initial value problem (1), (2) can be expressed in terms of operational calculus, but such expression seems to be practically useless for analysis. To obtain an approximate solution for the problem (1), (2) under the condition $\varepsilon \ll 1$ we used the method of averaging (by Krylov-Bogolyubov). The foundation of the method of averaging for integrodifferential equations with quasiperiodical kernels is a special mathematical problem. Some works by the authors of the present paper deal with this problem.

In the case when the frequency Ω coincides with one of the eigenfrequencies of internal waves the approximate solution for the problem (1), (2) takes the form

$$s^*(t, \varepsilon) = s_0 \cos \Omega t \cos(\sqrt{\varepsilon}\, \mu\, \Omega\, t); \quad \mu = \frac{1}{2}\,(2m + 1)^{-1} \quad (3)$$

The accuracy of the asymptotic solution in this case is

$$|s(t, \varepsilon) - s^*(t, \varepsilon)| \le c_1\sqrt{\varepsilon}\,, \qquad (0 \le t \le \frac{\theta}{\varepsilon})\ .$$

It follows from the expression (3) that in the case of resonance the periodical transfer of energy of oscillations takes place from the rigid body to internal waves and vise versa. This phenomenon was observed in a special experiment. A rectangular vessel made of organic glass was hung on strings. The vessel was half filled with water or another liquid. Then the other component was added, usually kerosene (sometimes there were other liquids). The depth of the layer for the second component was adjusted so that the resonance condition took place.

Having had been put out from the equilibrium position the system began to oscillate. We observed visually a gradual increase of the amplitude of internal waves and a decrease of oscillations of the vessel.

Due to viscosity, which any real liquid posesses, these beatings decayed (there are some real oscillograms).

If the internal resonance is absent, i.e. $\Omega \neq \omega_{2k+1}$, $k \neq$ $\neq 0, 1, \ldots$, then the calculations according to the averaging technique are more complicated. In this case the displacement of a rigid body with a cavity containing a heterogeneous liquid is approximately described by the formula

$$s^{*}(t, \varepsilon) = s_0 \cos \nu t; \qquad \nu = \Omega\left(1 + \frac{\varepsilon}{4} A\right),$$

$$A = \sum_{k=0}^{\infty} \frac{\omega^4_{2k+1}}{(2k + 1)^4} \frac{1}{\Omega^2(\Omega^2 - \omega^2_{2k+1})} .$$

The accuracy of the asymptotic solution in the nonresonant case is

$$|s(t, \varepsilon) - s^{*}(t, -\varepsilon)| \leq c_2\varepsilon , \qquad \left(0 \leq t \leq \frac{\theta}{\varepsilon}\right) .$$

The expression implies that in the nonresonant case the presence of internal waves causes only a small change of the oscillation frequency of a rigid body. The intensive exchange of energy is not observed.

It is easy to see experimentally the change of the frequency of free oscillations, if the vessel is firstly filled with a homogeneous liquid and then with a liquid containing two components.

Acknowledgement

All the experiments mentioned in this paper were carried out by N.L.Nikitin and L.Ju.Krasnopoljansky.

References

1. Sretenskii, L.N. Theory of ware motions of a fluid. Moscow, Nauka 1977 (In Russian).

2. Haskind, M.D. Hydrodynamical theory of oscillating motions of a ship. Moscow, Nauka 1972 (in Russian).

3. Akulenko,L.D.; Nesterov, S.V: Oscillations of a rigid body with a cavity containing a heavy inhomogeneous liquid. Mech. of Solids (Izv. AN SSSR. MTT) 21, No. 1 (1986) 26-35.

4. Akulenko, L.D.; Nesterov, S.V.: Non-resonant oscillations of a rigid body with a cavity containing a heavy two-layer fluid. Mech. of Solids (Izv. AN SSSR. MTT) 22, No. 2 (1987) 48-54.

5. Akulenko,L.D.: Kinematic control of the motion of a vessel with on ideal heavy fluid. Mech. of Solids (Izv. AN SSSR. MTT) 18, No. 1 (1983) 34-41.

6. Akulenko, L.D.; Nesterov, S.V.: Oscillations in a continuously stratified fluid in a moving vessel and their control. Appl. Math. and Mech. 51, No. 4 (1987) 456-462.

7. Akulenko, L.D.; Nesterov, S.V.: Oscillations of a solid at the interface between two fluids. Mech. of Solids (Izv. AN SSSR. MTT) 22, No. 5 (1987) 30-36.

Sensitivity Analysis for Structural Modeling and Design

N. V. BANICHUK, L. P. LUSCHIN, A. V. SHARANJUK

Institute for Problems in Mechanics
USSR Academy of Sciences, Moscow

SUMMARY

In this paper the approach to the correction of developed mathematical model is realized by means of the minimization of the functional of discrepancies between the frequencies and modes of free vibration measured in frequencies tests and estimated with help of the theoretical model. Modified model parameters (masses, rigidities and geometrical characteristics of the structural elements) are considered as unknown design variables. The weight constraints are imposed on these parameters. For effective solving of the discussed problem the procedure, based on application of adjoint variables and gradient algorithm, was proposed. For a numerical calculation of the discussed problem FORTRAN 77 program was created. This procedure and program have been used in numerical test for correction of mass-rigidity parameters for aeroplane wings of large aspect ratio.

Introduction

The objective of the paper is to establish a rational procedure for predicting the modified structure (mass and stiffnes matrices) which correlates most closely with a given set of measured modal frequencies. The sensitivity analysis method for effective calculation of basic functional variations depending on mass and rigidity variations is developed. The method is based on use of adjoint variables and adjoint system of equations. The technique allows to define the domains in the structure, where the remodelling seems to be most effective.

Note that a number of systematic procedures have been developed [1-2] and applied to specific problems. Other related approaches which modify analytical matrices based on a direct comparison of predictions and measurements have been published in [3]. A reviews of the previous results and existing procedures can be found in [4].

N. V. Banichuk · D. M. Klimov · W. Schiehlen (Eds.)
Dynamical Problems of Rigid-Elastic
Systems and Structures
IUTAM Symposium Moscow, USSR, May 23-27, 1990

1. Identification problem and application of sensitivity analysis.

Suppose that the model of the ideal structure is characterized by the vector $h(x)=\{h_1(x),h_2(x),...h_m(x)\}$ of design variables. Vertor $h(x)$ is defined for the domain Ω with boundary $\Gamma(x \in \Omega)$. The eigenfrequencies $\omega_i (\lambda_i=\omega_i^2)$ and the eigenmodes $u^i=u^i(x)=\{u_1^i(x),u_2^i(x),...,u_n^i(x)\}$ of free vibrations $(i=1,2,...)$ are the nontrivial solutions of the eigenvalue problem

$$Lu^i \equiv (C-\lambda_i A)u^i=0 \tag{1}$$

$$(Nu^i)_\Gamma=0 \tag{2}$$

Here L, N - operators of differentiation with respect to space variables, the coefficients of L, N depend analytically on h, i.e. $L=L(h)$, $N=N(h)$. Operators A and C are selfadjoint operators for boundary conditions (2) and, consequently, L is also the selfadjoint operator for considered boundary conditions.

Vector of design variables $h^*(x)=\{h_1^*(x),h_2^*(x),...,h_m^*(x)\}$, corresponding to actual structure is unknown beforehand and it is evaluation for large built up structure is difficult. The considered problem of deriving of the model for actual structure consists in determining the vector $h^*(x)$ using the identification methods.

Suppose that frequencies tests are accurately performed and permit to determine r eigenfrequencies and corresponding normal modes of free vibrations. By the same token the eigenvalues and eigenvectors λ_1^*, $\lambda_2^*,...,\lambda_r^*$; $u^{1*}(x),u^{2*}(x),...,u^{r*}(x)$ are supposed to be given for actual structure.

As a measure of the discrepancies between the characteristics of actual structure, corresponding the unknown vector $h^*(x)$, and characteristics obtained from (1), (2) for $h=h(x)$, we take the functional

$$J=\sum_{j=1}^{r}\{\alpha_j(\lambda_j-\lambda_j^*)^2+\beta_j(\hat{u}^j-\eta_j\hat{u}^{j*},\hat{u}^j-\eta_j\hat{u}^{j*})_\Omega\} \tag{3}$$

where the scalar products and normalized values are defined by the expressions

$$\hat{u}^j=u^j/(u^j,u^j)_\Omega^{1/2}, \quad (u^j,u^j)_\Omega=\int_\Omega u^ju^jd\Omega \tag{4}$$

Coefficients α_j, β_j - given nonnegative quantities (weight multipliers) which characterize the contribution of the corresponding vibration mode to the discrepancy functional J. The multipliers η_j in (3) can be chosen with the help of scalar product minimum condition. To produce the mathematical model of actual structure let us formulate the problem of finding the design variables vector, which minimize the discrepancy functional $J_*=\min_h J(h)$ and satisfies the system of integral constraints

$$J_i=\int f_i(h)d\Omega \quad (1,f_i)_\Omega=C_i, \; i=1,2,\dots,s \tag{5}$$

Here C_i - given constants, f_i - given functions. Superimposed design variables constraint (5) can include such isoperimetric conditions as constant volume condition. Let us vary the discrepancy functional. Performing the auxiliary transformations the expression for δJ can be written in the form

$$\delta J=\sum_{j=1}^{r}\{2\alpha_j(\lambda_j-\lambda_j^*)\delta\lambda_j+2\beta_j\Big((u^j,u^j)_\Omega^{-1/2}\delta u^j,\eta_j[(\hat{u}^j,\hat{u}^{j*})_\Omega\hat{u}^j-\hat{u}^{j*}]\Big)_\Omega+$$

$$+2\beta_j\delta\eta_j[\eta_j-(\hat{u}^j,\hat{u}^{j*})_\Omega] \tag{6}$$

We write the equations relating variations as terms in u^j, λ_i and h, which correspond to equations (1) and (2)

$$\delta(Lu^j)\equiv L(h)\delta u^j-\delta\lambda_jA(h)u^j+M(u^j,h)\delta h=0 \tag{7}$$

$$\delta(Nu^j)_\Gamma\equiv\big(N(h)\delta u^j+T(u^j,h)\delta h\big)_\Gamma=0 \tag{8}$$

Note that equations (7) and (8) depend linearly on $\delta u^j, \delta\lambda_j$ and δh. The variational equations (7), (8) are obtained from (1) and (2) by substituting $u^j+\delta u^j$, $\lambda_j+\delta\lambda_j$ and $h+\delta h$ for u^i, λ_j and h, respectively, and by isolating terms that are linear in δu^j, $\delta\lambda_j$ and δh. Here $M(u^j,h)$, $T(u^j,h)$ denote operators acting on the vector δh. We express the variation of the functional to be minimized in terms of the variation δh. To achieve that, we introduce an auxiliary vector function (the adjoint variable)
$V^j(x)=\{V_1^j(x),V_2^j(x),\dots,V_n^j(x)\}$, which we determine from the condition that the variation of the functional to be minimized should not contain the variation of the state variable δu^j. To accomplish this objective, it is necessary that V^j satisfy a certain system of differential equations and boundary conditions, i.e. we must define V^j as the solution of a certain boundary-value problem. Let us form the scalar product of the left-hand side of equation (7) and the vector V^j. Taking the integral over the domain Ω as the definition of this scalar product, we have

$$(V^j,L(h)\delta u^j)_\Omega-\delta\lambda_j(V^j,A(h)u^j)_\Omega+(V^j,M(u^j,h)\delta h)_\Omega=0 \tag{9}$$

Next, we perform integration by parts, transforming (9) in the equation

$$0=(V^j,L(h)\delta u^j-\delta\lambda_j Au^j+M(u^j,h)\delta h)_\Omega=(\delta u^j,LV^j)_\Omega-\delta\lambda_j(V^j,A(h)u^j)_\Omega+$$
$$+(\delta h,M^*(u^j,h)V^j)_\Omega \tag{10}$$

Some boundary terms that arise in the integration by parts are equal to zero because of equation (8), which is a consequence of the boundary condition of equation (2) and relates the boundary values of the variations δu^j and δh. The remaining boundary terms that arise are set equal to zero, requiring that the function V^j obey the boundary condition $(N(h)V^j)_\Gamma=0$. The coefficients of the adjoint operator M^* may depend on both h and u^j.

Let us derive formulas relating variations δh and variations of eigenfrequencies $\delta\lambda_j$. To this end multiply left hand side of the variation equation (7) by u^j and integrate the product over the domain Ω. Without offering detailed explanations, we point out the expression relating δh and $\delta\lambda_j$

$$\delta\lambda_j=(\delta h,R^j)_\Omega,\quad R^j\equiv M^*u^j/(u^j,Au^j)_\Omega \tag{11}$$

Taking into account (1), (3), (5), compose augmented Lagrange functional. Next, an expression for the variation of augmented functional δJ may be sought by varying the u^j, η_j, h and using the expressions (6), (10)

$$\delta J=\sum_{j=1}^{r}\{2\alpha_j(\lambda_j-\lambda_j^*)\delta\lambda_j+\left((u^j,u^j)_\Omega^{-1/2}\delta u^j,\eta_j[(\hat{u}^j,\hat{u}^{j*})_\Omega\hat{u}^j-\hat{u}^{j*}]\right)_\Omega+$$
$$+2\beta_j\delta\eta_j[\eta_j-(\hat{u}^j,\hat{u}^{j*})_\Omega]\}+\sum_{i=1}^{r}\left(\varkappa_i,\frac{\partial f_i}{\partial h}\delta h\right)+\left(\partial u^j,LV^j\right)_\Omega-$$
$$\delta\lambda_j(V^j,A(h)u^j)_\Omega+(\delta h,M^*(u^j,h)V^j)_\Omega \tag{12}$$

In the derivation of the relation between δJ variation δh we define the adjoint vectors V^j as a solutions of nonhomogeneous partial differential equations and boundary conditions

$$LV^j=2\beta_j\eta_j(u^j,u^j)_\Omega^{-1/2}[\hat{u}^{j*}-(\hat{u}^j,\hat{u}^{j*})_\Omega\hat{u}^j] \tag{13}$$

$$(NV^j)=0 \tag{14}$$

It is necessary for solvability of this problem that the right-hand side of equation (13) must be orthogonal to the eigenvectors of the corresponding homogeneous boundary value problem (1), (2). This requirement is satisfied for any values of the parameters $\beta_j\eta_j(\hat{u}^j,\hat{u}^{j*}-(\hat{u}^j,\hat{u}^{j*})_\Omega\hat{u}^j)_\Omega=0$.

A necessary condition for the extremum of J, regarded as a function of η_j, is given by the equality:

$$\eta_j=(\hat{u}^j,\hat{u}^{j*})_\Omega \tag{15}$$

The multipliers η_j, defined according to (15), satisfy inequality $\eta_j\le 1$. The equality $\eta_j=1$ is attained if the vectors $\hat{u}^j(x)$ and $\hat{u}^{j*}(x)$ are collinear for every point $x\in\Omega$. Take into

account the expressions (11), (15) for variations of eigenvalues $\delta\lambda_j$ and for multipliers η_j and use the equation (13) for adjoint variables V^j.
As a result we have the following expression for δJ

$$\delta J=\Big(\delta h,\ \sum_{j=1}^{r}\{M^*V^j+R^j[2\alpha_j(\lambda_j-\lambda_j^*)+(V^j,Au^j)_\Omega]\}+\sum_{i=1}^{s}æ_i\ \frac{\partial f_i}{\partial h}\ \Big)_\Omega \qquad (16)$$

Formula (16) determines the sensitivity of the optimized functional to the variations of model parameters h and gives us the possibility to develop numerical methods of successive optimization. Various methods were used within the successive optimmization algorithm, but the gradient method appears to be the most efficient from a practical point of view.

Mass and stiffness identification

As an example, let us give the solution for the problem of mass and stiffness identification for the wing of large aspect ratio. It is very important problem, for example, on glider, which in many cases can not be completely located in flutter experiments in an aerodynamic wind tunnel because of high aspect ratio. In this cases only accurate mathematical model can yield accurate speed of bending-torsion wing flutter.

For example, considere real wing of one aeroplane with mass-rigidity characteristics distributed over the span. Dynamically similar wing consisted of 8 sections and point load at the free end of the wing. For this model the frequency tests were performed and the frequencies and modes of the lower three tones of bending-torsion vibrations were measured.
For computation of vibrational modes and frequencies the input data were defined for the 20 sections of the wing and the beam-like wing model was used. The result of computations of vibrational eigenfrequencies are brought in table 3 for two cases of the input date assignment.

Table 1

Case	$\omega_1(\frac{rad}{s})$	$\omega_2(\frac{rad}{s})$	$\omega_2(\frac{rad}{s})$
a)	12.9	46.6	49.2
b)	12.9	46.3	49.0
c)	11.52	42.9	48.6
d)	11.5	42.1	48.2

In the first case (case a) the input date were defined as the step-functions, i.e. input date are constant along the wing boxesand very step-wise when we transfer from one part of the wing to another. In the second case (case b) the input data were approximated by piece-wise linear functions (c - frequency test, d - after correction). The results of the computation show that the input data affect on the frequencies of the vibrations (table 1) and vibrational modes. Both in case a) and in case b) computed the second and the third modes of bending-torsion vibrations differ from the measured vibrational modes of wing (fig. 2). For correction of the parameters of the mathematical model (input data correction) we minimize the discrepancy between computed eigenfrequencies and eigenmodes of vibrations and eigenfrequencies and eigenmodes measured in frequency test. As a result the corrected mass-rigidity input data were determined for dynamically similar wing.[*)] This distribution (c) differs insignificantly from the approximation (a) and (b), see fig. 1. Computations, performed with application of corrected data, reduced to satisfactory coincidence of computed eigenfrequencies and eigenmodes of vibrations and the corresponding values measured in frequency tests (see fig.2)

[*)] The computations were performed with application of described procedure and developed package of programs "REZALT" that consists of 2000 operators on FORTRAN-77 language and keep in Institute for Problems in Mechanics USSR Academy of Sciences and Central Aero-Hydrodynamic Institute of USSR. We are ready to offer this program for all who needs it.

Table 2

Case	Flutter speed $(\frac{m}{s})$	$\omega_{cr}(\frac{rad}{s})$
a)	314.2	39.8
b)	296.0	40.0
c)	245.9	36.6

Table 2 represent computed velocities and frequencies of bending-torsion flutter for the next three cases when the input date were given by step-functions (a), were approximated by piece-wise linear function (b), were obtained as a result of the correction (c). It is obvious from the comparison of the results presented in table 2, that the computed velocities and frequencies of bending-torsion flutter depend essentially on input data.

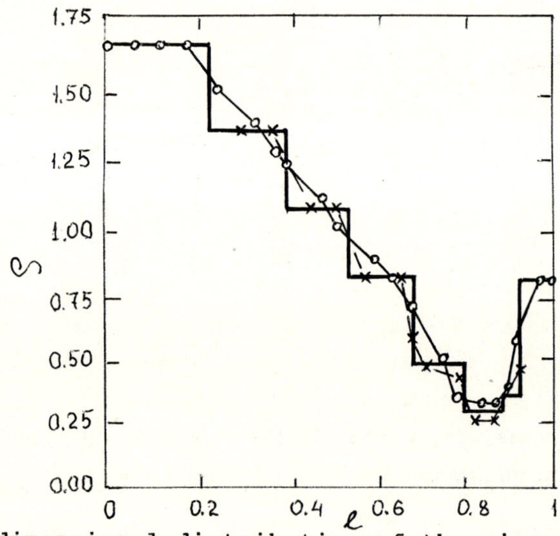

Fig. 1. Nondimensional distribution of the wing mass.

a) ———————— step-functions distribution of the wing mass

b) ——o——o—— linear approximation of the wing mass

c) ——x——x——x calculated distribution of the wing mass after correction.

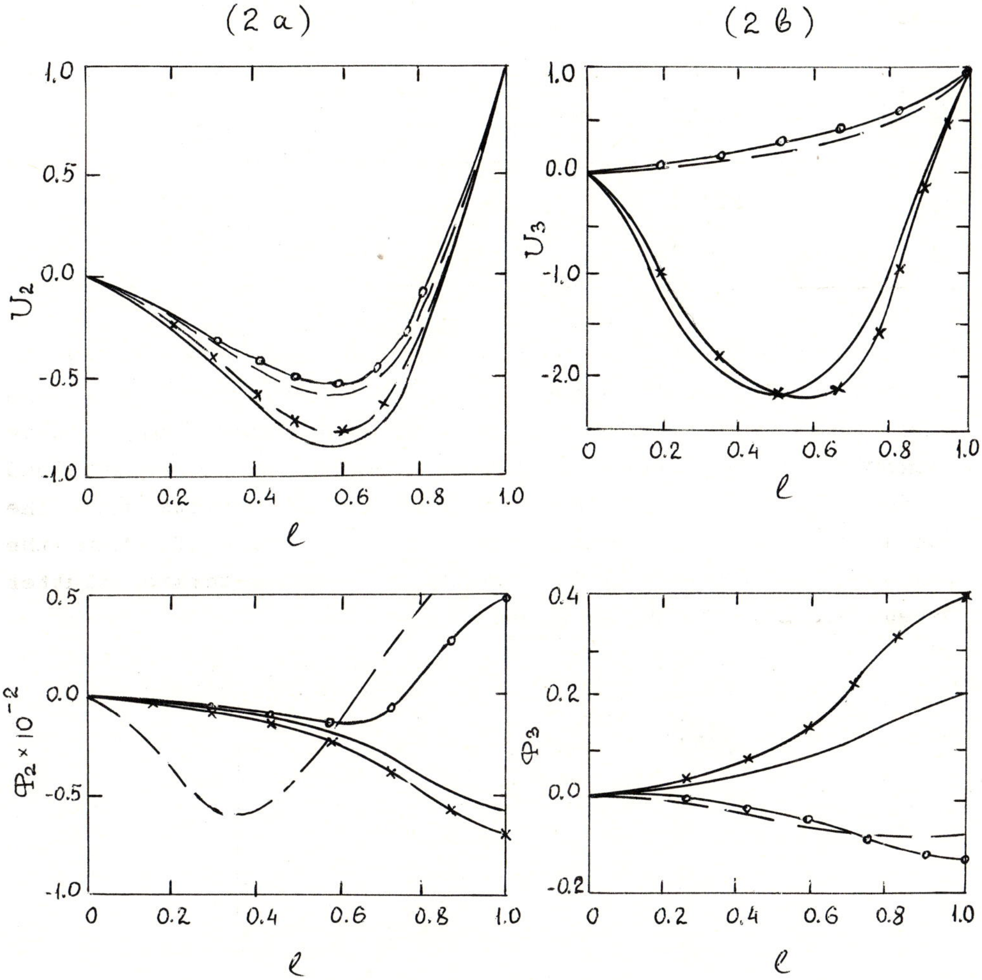

Fig. 2. Second (2a) and third (2b) modes of bending-torsion vibrations of the wing.

a) ——————— modes measured in frequency test

b) — — — — modes calculated for piece-wise (step-function) approximation

c) —o—o—o— modes calculated for piece-wise linear approximation

d) —x—x—x— modes calculated after correction

In this case only accurate mathematical model after correction without mistakes in mass and elastic properties of wing can yield accurate speed of bending-torsion glider wing flutter and assign V_D(design dive speed). Each percent of precision in calculation of bending-torsion flutter speed is a problem of big importance for wing glider because of the fact that more accurate V_D one of the main performance parameters would give a new opportunity for the world leadership.

References

1. Rodden W.P. "A method for deriving structural influence coefficients from ground vibration tests", AIAA J., vol. 5, n° 5, 1967, p. 991-1000.

2. Taylor J.E. "Scaling a discrete structural model to match measured modal frequencies", AIAA J., vol. 15, n° 11, 1977, p. 1647-1649.

3. Banichuk N.V., Luschin L.P., Makeev E.V., Sharanyuk A.V. "Modeling and design of complex structures" (in Russian) Proc. of Gagarin scientific seminars on cosmonautics and aviation (in Russian), Nauka, Moscow, 1987, p. 177-179.

4. Red'ko S.F., Ushkalov V.F., Yakovlev V.P. "Identification of mechanical systems" (in Russian), Naukova Dumka, Kiev, 1985.

Dynamics of Space Systems Including Elastic Tethers

V. V. BELETSKY

Keldysh Institute of Applied Mathematics
the USSR Academy of Sciences
Moscow

E. M. LEVIN

Institute for Mechanical Engineering Research
the USSR Academy of Sciences
Moscow

Summary. Stability of stationary motions of an orbital system, including two satellites connected by an elastic massive tether, is studied. Radial equilibrium configurations of the system in the gravitational field are stable. Aero- and electrodynamic forces turn to be responsible for the flutter-type instability of inclined configurations. Internal friction introduces natural damping into the system. Certain restrictions on operational characteristics of the system arise from the stability conditions.

1. Introduction. After an impressive flow of innovative ideas and consequent profound studies by many investigators, space tethers have been eventually recognized as a highly promising new space technology. The first mission of the Shuttle-based Tethered Satellite System (TSS) is scheduled by NASA for 1991. During this flight a half-ton subsatellite will be deployed upwards from the Shuttle Orbiter on a 20 km conducting tether to carry out electrodynamic experiments. During the second TSS mission the subsatellite will be deployed Earthwards to the altitude of 120 km to accomplish aerodynamic measurements.

The effect of electro- and aerodynamic forces on the TSS stability will be the main problem under consideration in this study.

2. Equations of motion. Let the Shuttle Orbiter follow a circular orbit. Motion of the subsatellite A relative to the orbital frame attached to the center of mass O of the Orbiter is governed by the ordinary differential equation

$$m_A\left[\ddot{\mathbf{r}}_A + 2\mathbf{\Omega}\times\dot{\mathbf{r}}_A + \mathbf{\Omega}\times(\mathbf{\Omega}\times\mathbf{R}_A) + \mu\mathbf{R}_A R_A^{-3}\right] = -\mathbf{T}_A + \mathbf{F}_A, \qquad (1)$$

while motion of a massive and extensible tether is determined by the partial differential equations

$$\begin{gathered}\rho\left[\ddot{\mathbf{r}} + 2\mathbf{\Omega}\times\dot{\mathbf{r}} + \mathbf{\Omega}\times(\mathbf{\Omega}\times\mathbf{R}) + \mu\mathbf{R}R^{-3}\right] = \mathbf{T}' + \mathbf{F},\\ \mathbf{T} = \mathbf{e}E(\gamma-1).\end{gathered} \qquad (2)$$

N.V. Banichuk · D. M. Klimov · W. Schiehlen (Eds.)
Dynamical Problems of Rigid-Elastic
Systems and Structures
IUTAM Symposium Moscow, USSR, May 23-27, 1990

Here $\mathbf{r}$ is the radius vector in the orbital axes, $\mathbf{R}$ the geocentric radius vector, $\mathbf{\Omega}$ the orbital angular rate, $\mathbf{T}$ the tether tension force, E the axial stiffness of the tether, ρ the mass per unit length of the tether, m_A the subsatellite mass, μ the Earth's gravity constant, $\mathbf{F}$ and $\mathbf{F}_A$ represent disturbing forces, dots indicate differentiation with respect to time, while primes denote differentiation with respect to a natural parameter s measured from the Shuttle towards the subsatellite, $0 \leqslant s \leqslant l$, l is the total length of the unstressed tether, $\gamma = |\mathbf{r}'|$ the elongation of the tether, $\mathbf{e} = \mathbf{r}'/\gamma$ the unit tangent vector.

The tether supporting the subsatellite is attached to the Shuttle, that means

$$\mathbf{r}_0 = 0. \tag{3}$$

3. Stability of the radial equilibrium. If no perturbations ($\mathbf{F} = 0$, $\mathbf{F}_A = 0$), the TSS will have two equilibrium positions relative to the orbital frame: one, with the tether stretched upwards, and the second, with the tether deployed downwards. Equilibria with a slack tether are excluded since they proved to be unstable.

For $\mathbf{F} = 0$, $\mathbf{F}_A = 0$, equations admit of the energy integral

$$K + W = h \tag{4}$$

where

$$\begin{aligned} K &= \frac{1}{2} m_A \dot{\mathbf{r}}_A^2 + \frac{1}{2}\int_0^l \rho \dot{\mathbf{r}}^2 \, ds, \\ W &= \frac{1}{2}\int_0^l E(\gamma - 1)^2 \, ds - \int_0^l \rho \left[\frac{1}{2}(\mathbf{\Omega} \times \mathbf{R})^2 + \frac{\mu}{R}\right] ds - \\ &\qquad m_A \left[\frac{1}{2}(\mathbf{\Omega} \times \mathbf{R}_A)^2 + \frac{\mu}{R_A}\right]. \end{aligned}$$

The potential energy W was found to be positive definite in the vicinity of the radial equilibrium under the condition

$$E > E_* \approx T_A \tag{5}$$

where T_A is the equilibrium tension at the end A. Actually, we always have $E \gg T_A$ since an acceptable strain of any real tether is rather small: $\varepsilon = \gamma - 1 = T/E \ll 1$. Therefore, the radial equilibrium is stable in the Lyapunov's sense.

4. Natural modes of in-plane oscillations. Small oscillations around the radial equilibrium in the orbital plane are governed by the equations

$$\begin{aligned} &\rho(\delta\ddot{x} - 2\Omega\delta\dot{y} - 3\Omega^2\delta x) = (E\delta x')', \\ &\rho(\delta\ddot{y} + 2\Omega\delta\dot{x}) = (T\gamma^{-1}\delta y')', \\ &m_A(\delta\ddot{x}_A - 2\Omega\delta\dot{y}_A - 3\Omega^2\delta x_A) = -(E\delta x')_A, \\ &m_A(\delta\ddot{y}_A + 2\Omega\delta\dot{x}_A) = -(T\gamma^{-1}\delta y')_A, \\ &\delta x_0 = 0, \quad \delta y_0 = 0, \end{aligned} \tag{6}$$

where the functions $T = T(s)$ and $\gamma = T(s)E^{-1}$ describe the equilibrium, and δx, δy represent longitudinal and transverse displacements. While deriving (6), some terms of the order of $l/R \ll 1$ were neglected.

By setting $\delta x = U(s)\cos\beta t$, $\delta y = V(s)\sin\beta t$, the appropriate eigenvalue problem is obtained from (6). If we omit the giroscopic terms, longitudinal and transverse oscillations will become decoupled. Then, for longitudinal oscillations we get

$$U_n(s) = \sin(k_n s), \quad \beta_{en}^2 = k_n^2 v_e^2 - 3\Omega^2, \quad n = 1, 2, \ldots, \tag{7}$$

where $v_e = \sqrt{E/\rho}$ is the longitudinal wave velocity and the wave numbers k_n represent the positive solutions of the frequency equation $(kl)\tan(kl) = m_t/m_A$, m_t being the tether mass. For transverse oscillations, we derive

$$V_n(s) = \frac{1}{2}\left[P_\nu\left(\frac{s}{L_0}\right) - P_\nu\left(-\frac{s}{L_0}\right)\right], \quad \beta_{tn}^2 = \frac{3}{2}\Omega^2\nu(\nu+1), \tag{8}$$

where P_ν is the Legendre function of the first kind and ν are the positive solutions of the frequency equation

$$2\xi\frac{d}{d\xi}\left[P_\nu(\xi) - P_\nu(-\xi)\right] = \nu(\nu+1)\left[P_\nu(\xi) - P_\nu(-\xi)\right].$$

Here $\xi = lL_0^{-1}$, $L_0 = \sqrt{2T_0/3\rho\Omega^2}$, and T_0 is the equilibrium tension at the Shuttle end. In particular, there always exists pendulum libration mode $\nu = 1$, $V_1 = sL_0^{-1}$, $\beta_{t1} = \sqrt{3}\Omega$.

The giroscopic terms cause a weak coupling between longitudinal and transverse oscillations which is very significant for the behavior of the system. In the leading terms of a series expansion in powers of the small parameter $\Omega l/v_e \ll 1$, the high frequency longitudinal oscillations (7) generate the weak transverse response

$$V_n(s) = -2\Omega\beta_{en}^{-1}U_n(s), \tag{9}$$

while the low frequency transverse oscillations (8) produce the weak longitudinal response

$$U_n(s) = 2\Omega\beta_{tn}\int_0^s \frac{d\eta}{E}\left[m_A V_n(l) + \int_\eta^l \rho V_n(\xi)\,d\xi\right]. \tag{10}$$

It is appropriate to call the oscillations (7), (9) *predominantly longitudinal* and the oscillations (8), (10) *predominantly transverse*.

5. The effect of small perturbing forces. Let us introduce now vertical (F_x) and horizontal (F_y) components of small perturbing forces into the right hand parts of the equations (6). In the steady state case, they can be represented as linear functions of the displacements δx, δy and their first derivatives

$$\begin{aligned} F_x &= \alpha(a_0 + a_1\delta x + a_2\delta y + a_3\delta x' + a_4\delta y' + a_5\delta\dot{x} + a_6\delta\dot{y}), \\ F_y &= \alpha(b_0 + b_1\delta x + b_2\delta y + b_3\delta x' + b_4\delta y' + b_5\delta\dot{x} + b_6\delta\dot{y}), \end{aligned} \tag{11}$$

where α is a small parameter. Then, pumping or dissipation of the energy (4) will be determined by the work of the perturbing forces

$$\begin{aligned} &h(t+\Delta t) - h(t) = \\ &\int_t^{\Delta t}\int_0^l (F_x\delta\dot{x} + F_y\delta\dot{y})\, ds dt + \int_t^{\Delta t} (F_{Ax}\delta\dot{x}_A + F_{Ay}\delta\dot{y}_A)\, dt. \end{aligned} \tag{12}$$

Trajectories of natural oscillations in the perturbed system are defined as

$$\begin{pmatrix}\delta x\\ \delta y\end{pmatrix} = ce^{\varepsilon_n t}\left[\cos\beta_n t\begin{pmatrix}U_n^c(s)\\ V_n^c(s)\end{pmatrix} + \sin\beta_n t\begin{pmatrix}U_n^s(s)\\ V_n^s(s)\end{pmatrix}\right] \tag{13}$$

where c_n are constant, $\lambda_n = \varepsilon_n + i\beta_n$ are the eigenvalues and U_n^c, U_n^s, V_n^c, V_n^s the eigenfunctions of the perturbed problem. When $\alpha \to 0$, the real parts of the eigenvalues $\varepsilon_n \to 0$, the functions $U_n^s \to 0$ and $V_n^c \to 0$, while the imaginary parts β_n and functions U_n^c, V_n^s transform into the natural frequencies and modes of the coupled longitudinal-transverse oscillations (7)–(10).

In order to find ε_n in a linear approximation in α, it is necessary to calculate the work of perturbing forces along the generating trajectories of oscillations. Over a period of oscillations $\Delta t_n = 2\pi/\beta_n$ according to (12), (13)

$$h(t + \Delta t_n) = h(t)\exp(2\varepsilon_n \Delta t_n) \approx h(t) + 2\varepsilon_n \Delta t_n h(t).$$

Writing down and transforming an expression for the energy $h(t)$, we obtain finally the asymptotic representation

$$\varepsilon_n = \frac{W_{tn} + W_{An}}{H_{tn} + H_{An}} + O(\alpha^2) \tag{14}$$

where

$$\begin{aligned} W_{tn} &= \frac{1}{\Delta t_n}\int_0^{\Delta t_n}\int_0^l (F_x\delta\dot{x} + F_y\delta\dot{y})_n\, ds\, dt, \\ W_{An} &= \frac{1}{\Delta t_n}\int_0^{\Delta t_n} (F_{Ax}\delta\dot{x}_A + F_{Ay}\delta\dot{y}_A)_n\, dt, \\ H_{tn} &= \int_0^l \rho(\beta_n^2 U_n^2 + \beta_n^2 V_n^2 + 2\Omega\beta_n U_n V_n)\, ds, \\ H_{An} &= m_A(\beta_n^2 U_n^2 + \beta_n^2 V_n^2 + 2\Omega\beta_n U_n V_n)_A, \end{aligned}$$

and the work of perturbing forces is calculated along the generating trajectories $\delta x = U_n(s)\cos\beta_n t$, $\delta y = V_n(s)\sin\beta_n t$ which is indicated by the subscript "n".

For the case when perturbing forces are not small, appropriate numerical procedures were developed to compute the eigenvalues. The analytical and numerical results were found to be in accordance and are presented further.

6. The effect of aerodynamics. To help appreciation of the system dynamics, we shall simplify the problem and ignore rotation and oblateness of the atmosphere. By applying the technique of analytic calculations in the form (14), asymptotic stability of all the modes of oscillations, except for the lowest predominantly transverse one, was established. This result is apparently attributed to the aerodynamic friction. For the higher modes of the predominantly transverse oscillations, the decrements are given by

$$\varepsilon_{tn} = -\left(\varkappa_A V_{nA}^2 + \int_0^l \varkappa V_n^2\, ds\right)\left(m_A V_{nA}^2 + \int_0^l \rho V_n^2\, ds\right)^{-1} \tag{15}$$

where $n = 2, 3, \ldots$, $\varkappa_A = \frac{1}{2} c_A \Sigma_A \rho_{aA} v_A$, $\varkappa = \frac{1}{2} c_t d_t \rho_a v$, c_A and c_t being the air drag coefficients of the subsatellite and tether, respectively, ρ_a the air density, v the orbital velocity, d_t the tether diameter, Σ_A the subsatellite cross section area.

The expressions for the real parts of eigenvalues ε_{en}, referring to the predominantly longitudinal oscillations, are very much alike, although more lengthy. Actual rates of decay turn to be very low. Say, for a considerably low subsatellite altitude of 110 km, amplitudes of oscillations diminish only by several percent per orbit. The same concerns to all modes of out-of-plane oscillations.

The situation with the lowest mode of the predominantely transverse oscillations is quite different. The expression for the real part of the corresponding eigenvalue contains additional terms compared to the plain structure of (15):

$$\begin{aligned}\varepsilon_{t1} = \frac{1}{2}\rho_{aA} v_A \big[&-(c_A \Sigma_A + c_t d_t H_a)(m_A + m_t/3)^{-1} + \\ &\Omega l v_A (c_A \Sigma_A + c_t d_t H_\rho)(E H_\rho)^{-1}\big].\end{aligned} \tag{16}$$

Here, $H_a = k\Theta p^{-1}$ is the so-called "height of a uniform atmosphere", k being the Boltzmann's constant, Θ the temperature of the air, and p the average weight of molecules at the subsatellite altitude, while H_ρ represents a somewhat different characteristics of the air density profile

$$H_\rho = -\rho_a \left(\frac{d\rho_a}{dh}\right)^{-1} \approx H_a \left(1 + \frac{dH_a}{dh}\right)^{-1}. \tag{17}$$

The asymptotic stability condition can now take the form of the restriction on the axial tether stiffness

$$E > E_{**} \approx \frac{(m_A + m_t/3)\Omega l v_A}{H_\rho} \cdot \frac{c_A \Sigma_A + c_t d_t H_\rho}{c_A \Sigma_A + c_t d_t H_a}. \tag{18}$$

Notice that it appears to be independent of the air density ρ_a. Therefore and surprisingly, a massive subsatellite (large m_A) on a thin tether (small E) will never gain stability by being withdrawn from dense atmospheric layers. The only satisfactory result will be that feebler aerodynamic forces will produce smaller instability increments since ε_{t1} is proportional to ρ_a.

An appropriate examination of the matter has shown that the energy pumping comes through aerogradient components of the air drag forces, *i.e.* through the increments of the horizontal forces F_x caused by the vertical displacements δy and the corresponding changes in the air density ρ_a, acting along the trajectories of oscillations at the lowest frequency $\beta_{t1} \approx \sqrt{3}\Omega$ as shown in Fig. 1(b). If no giroscopic coupling between longitudinal and transverse oscillations which is inherent in any relative motion in orbit, no instability of this kind emerges.

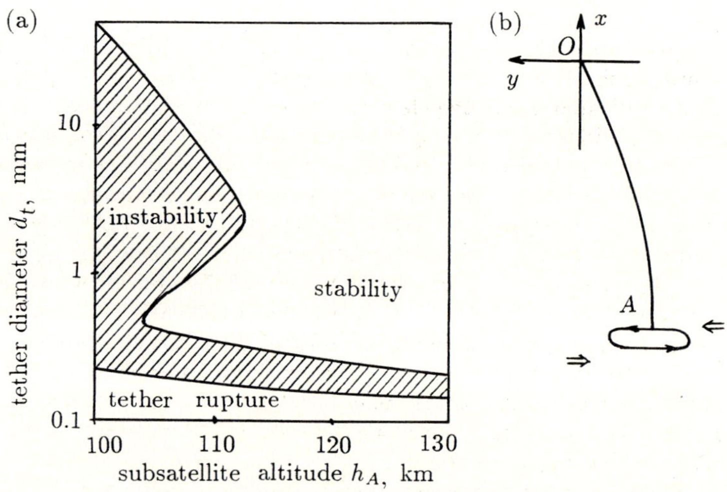

Fig. 1. (a) The region of stability of an atmospheric tethered subsatellite has a curious "nose" which is difficult to penetrate deeper into the atmosphere below a certain critical altitude of around 100 km. (b) Energy is pumped through the lowest mode of the coupled transverse-longitudinal oscillations.

As a result of vast investigations covering a wide range of system parameters, a typical location of stability region on the plane of two significant parameters, the subsatellite altitude h_A and the tether diameter d_t, is indicated in Fig. 1(a). The lower boundary of the stability region lies quite as predicted by

(18). But in the vicinity of 100 km altitude it gets twisted forming a nose-like protrusion. This is attributed to a considerable curvature of the tether line that is characteristic of its steady state configurations in dense atmospheric layers [1,2]. The point is, as it was concieved, that the coupling between longitudinal and transverse oscillations is much stronger for a noticebly curved tether than for a straight one, and the longitudinal response to transverse oscillations is of far greater magnitude than that of (10). Respectively, the energy pumping by the aerogradient components of aerodynamic forces becomes far more intensive and overcomes the aerodynamic friction.

Still, we can utilize the once derived formula (18) to get a generally good approximation by substituting for the physical stiffness E an "effective stiffness" of curved tether E_{eff}. The latter has an explicit representation to be given further.

As it is clear from Fig. 1(a), stable flight of the subsatellite below the critical altitude of about 100 km is possible only when using an unreasonably thick tether.

7. The effect of electrodynamics. To simplify the problem, assume that the axis of the Earth's magnet is coincident with the axis of the Earth's rotation. As indicated in [3], the electrodynamic tether system has four equilibrium configurations when moving in the equatorial plane. This variety arises from the following possibilities: firstly, the tether can be deployed upwards or downwards, and secondly, the system can operate in two modes producing either electricity power, or low thrust. When in the thrust mode, the tether is slanted towards the flight direction under the action of Amper forces (configurations OA_2 and OA_3 in Fig. 2(a)), while during power generation it is inclined in the opposite direction (configurations OA_1 and OA_4).

Results of both analytical and numerical investigations of stability are qiute disappointing: *no one* of the above mentioned equilibria is stable if the current in the tether is constant. The energy pumping is attributed to nonconservative components of Amper forces acting along the trajectories of coupled transverse-longitudinal oscillations as shown in Fig. 2(a).

For the lowest mode of the predominantly transverse oscillations, it can be obtained with the help of (14) that

$$\varepsilon_{t1} = \Omega IBlE^{-1} \tag{19}$$

where I is the current in the tether and B the geomagnetic induction. This formula proves to be valid only for sufficiently small currents. For relatively big currents it needs adjustment by replacing the physical stiffness E by the *effective stiffness*

$$E_{eff} = lJ_0\left(J_0J_2 - J_1^2\right)^{-1} \tag{20}$$

where

$$J_k = \int_0^l \frac{\sin^k\theta}{T\cos^2\theta}\,ds, \qquad k = 0, 1, 2,$$

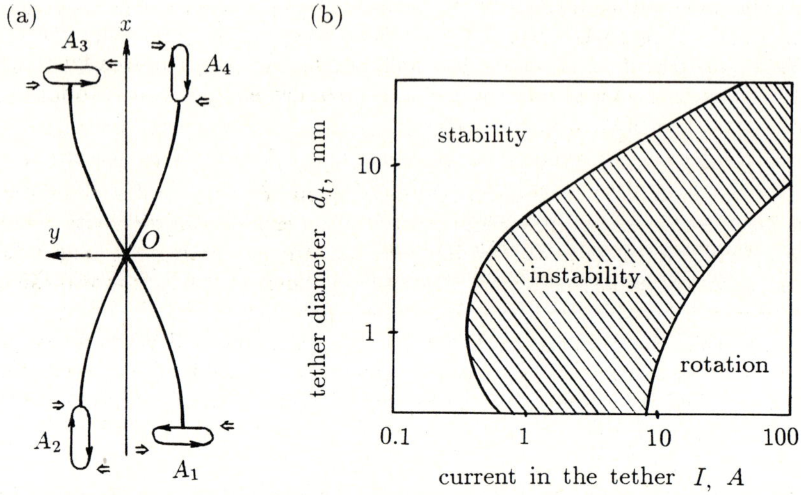

Fig. 2. (a) For different equilibrium configurations energy is pumped through different modes. (b) Damping due to internal friction in the tether can improve the situation though it is incapable of maintaining stability in the whole region $I < I_*$ where the equilibria exist.

and the functions $T = T(s)$, $\theta = \theta(s)$ describe equilibrium configuration, T being the tether tension, θ the inclination to the local vertical.

Here, the effective stiffness is defined as follows. Suppose that by applying a small constant force dT along the tether line at the end A, an equilibrium configuration OA is transformed into a close one OA'. Then, let $E_{eff} = l(dT/du)$ where du is the displacement AA'. For a straight tether by this definition $E_{eff} = E$, as expected. But for a curved tether, the result will be different as the displacement AA' is partly connected with alteration of the tether curvature. Thus, even an inextensible tether when curved will possess a certain effective stiffness. The same concerns to the atmospheric tether system discussed above. Note that this quasi-static definition suits well *only* for the lowest quasi-pendulum mode, but not for the higher ones.

In the present case, $E_{eff} \approx k^{-1} T_0 I_*^2 I^{-2}$ where T_0 is the equilibrium tension at the Shuttle end, $k \approx 0.1$, $I_* \approx 3(m_A + m_t/2)\Omega^2 B^{-1}$ is the maximum current before the system comes into rotation [3]. Therefore, for relatively big currents the instability increment grows rather rapidly, as I to the third power:

$$\varepsilon_{t1} \approx \Omega k I^3 I_*^{-3}. \tag{21}$$

This is completely confirmed by numerical investigations. The instability becomes fairly strong when the current I is comparable to the critical one I_*.

When the current I is relatively small, stability can be acquired by varying I according to a special control law [3]. A similar effect can be produced by natural internal friction in the tether.

8. The effect of internal friction. As reported by many authors of experimental studies, intensive internal dissipation of energy is characteristic of braided tethers which is due to friction between neighbouring elementary fibers during oscillations. There is a little doubt that the predominantly longitudinal oscillations will experience vigorous damping. Once excited, they will die out in minutes or even in seconds as by the easiest estimate.

Unlike this, the predominantly transverse oscillations will be damped quite unsatisfactory since coupling between transverse and longitudinal oscillations is rather faint. By applying the technique of asymptotic calculations, we find the decrements to be

$$\varepsilon_{tn} = -\frac{2\eta\Omega^2}{E\beta_{tn}^3} \cdot \frac{\int_0^l (TV_n')^2 \, ds}{m_A V_{nA}^2 + \int_0^l \rho V_n^2 \, ds} \tag{22}$$

where η is the loss factor that can be up to 0.1. According to this formula, a typical rate of amplitude's decay for the lowest mode of oscillations will be at best up to a percent per orbit. The higher the mode, the feebler damping is predicted by (22).

Nevertheless, it is enough to provide stability of electrodynamic tether systems for relatively small currents as indicated in Fig. 2(b).

9. The main points.

- The radial equilibrium of the tethered subsatellite in the gravitational field is stable in the Lyapunov's sence
- There is giroscopic coupling between longitudinal and transverse in-plane oscillations, and two types of natural modes can be distinguished: predominantly-transverse and predominantly longitudinal
- Nonconservative components of the aero- and electrodynamic forces are capable of pumping energy into certain modes of coupled transverse-longitudinal oscillations thus destroying the stability
- Tether curvature enlarges energy pumping since the coupling becomes stronger; the notion of effective longitudinal stiffness of curved tether help to appreciate this effect
- Natural damping in the system is represented by aerodynamic friction and internal friction in the tether and provide stability only in certain parts of desired operational regions
- Trying to penetrate deeper into the atmosphere or gain higher levels of electricity or thrust generation, a strong flutter-type instability is met that urges to seek augmented active control over the system

References

1. V. V. Beletsky and E. M. Levin, *Mechanics of an Orbital Cable System*, Cosmic Research **18** (1980), 486–494.
2. E. M. Levin, *Stability of the Time-Independent Tethered Motions of Two Bodies in Orbit Under the Action of Gravitational and Aerodynamic Forces*, Cosmic Research **22** (1984), 544–551.
3. E. M. Levin, *Stability of the Stationary Motions of an Electromagnetic Tether System in Orbit*, Cosmic Research **25** (1987), 368–376.

Variational Approaches for Dynamics: Numerical Studies

M. Borri† F. Mello‡ S. N. Atluri‡

†Dipartimento di Ingegneria Aerospaziale
Politecnico di Milano
Milano 20133 Italy
‡Center for Computational Mechanics
School of Civil Engineering
Georgia Institute of Technology
Atlanta, GA 30332

1 Abstract

Weak formulations in Analytical Dynamics are developed, paralleling the variational methods in elastostatics, and including a fundamental yet novel approach for treating constraints (both holonomic and nonholonomic). A general three field approach is presented, in which the momentum balance conditions, the compatibility conditions between displacement and velocity, the constitutive relations and the displacement and momentum boundary conditions are all enforced in weak form. A primal or kinematic formulation is developed from the general form by enforcing the compatibility conditions and displacement boundary conditions *a priori*. The conditonal stability of the kinematic formulation is the counterpart of locking phenomenon in elastostatics and may be avoided either by reduced order integration or by a mixed formulation. Toward this end, a two field mixed formulation is presented, which follows from the general form when the constitutive relations are satisfied *a priori*. A general set of the constraint equations are introduced into the kinematic and mixed formulations using a specific choice of multipliers which results in a modified variational principles. Two simple examples concerning rigid body dynamics are presented.

2 Introduction

Recently there has been a renewed interest in the study of multi-body dynamics and its application to a wide variety of engineering problems. The nonlinear equations of motion, in explicit form are quite complex due to the expression for absolute acceleration. These complexities are avoided if a weak formulation is employed. The priciple of virtual work, or Hamilton's principle is one such weak formulation. There has been much discussion in the literature [1] [2] [3], concerning Hamilton's principle as a starting point for the numerical solution of dynamics

N. V. Banichuk · D. M. Klimov · W. Schiehlen (Eds.)
Dynamical Problems of Rigid-Elastic
Systems and Structures
IUTAM Symposium Moscow, USSR, May 23-27, 1990

problems. Much of this discussion involves the conditions under which Hamilton's principle may be stated as the stationarity condition of a scalar functional. Due to the unsymmetric character of initial value problems, the governing equations are not expressible as such a condition. This fact in no way deminishes the usefullness of variational approaches for initial value problems. In fact, drawing on the mature literature concerning variational methods in solid mechanics, very general weak forms can be developed for dynamic systems, the most general being analogous to a Hu - Washizu type formulation. The principle of virtual work or Hamilton's principle is obtainable from the general weak form by satisfying displacement compatibility and the displacement boundary conditions *a priori* . A Hamiltonian or complementary energy approach is obtained by satisfying the constitutive relations between momentum and velocity *a priori* .

In this paper several formulations, for Analytical Dynamics are discussed, with the objective of developing a system of equations which may be directly implemented in the framework of time finite elements. This approach leads to a set of nonlinear equations, which are solved using Newton's method. Several simple example concerning rigid body dynamics are presented to illustrate the feasibility of the methods.

3 Weak Formulations in Analytical Dynamics

Let us consider a dynamical system with generalized coordinates $\boldsymbol{q}$, acted upon by a generalized external load $\boldsymbol{f}$. We denote the Lagrangian of the system by $\mathcal{L}$, regarded as a function of the coordinates, $\boldsymbol{q}$, the velocities $\boldsymbol{v}$ and time t.
The equations of motion:

$$\boldsymbol{f} + \frac{\partial \mathcal{L}}{\partial \boldsymbol{q}} - \dot{\boldsymbol{p}} = 0 \tag{3.1}$$

the conditions of compatibility and the constitutive relations:

$$\boldsymbol{v} = \dot{\boldsymbol{q}} \qquad \boldsymbol{p} = \frac{\partial \mathcal{L}}{\partial \boldsymbol{v}} \tag{3.2}$$

the displacement and momentum boundary conditions:

$$\boldsymbol{q}(t_k) = \boldsymbol{q}_k \qquad \boldsymbol{p}(t_k) = \boldsymbol{p}_k \qquad k = 1, 2 \tag{3.3}$$

can be combined in the following weak form:

$$\begin{aligned} \int_{t_1}^{t_2} \left\{ \left(\boldsymbol{f} + \frac{\partial \mathcal{L}}{\partial \boldsymbol{q}} - \dot{\boldsymbol{p}} \right) \cdot \delta \boldsymbol{q} - (\boldsymbol{v} - \dot{\boldsymbol{q}}) \cdot \delta \boldsymbol{p} - \left(\boldsymbol{p} - \frac{\partial \mathcal{L}}{\partial \boldsymbol{v}} \right) \cdot \delta \boldsymbol{v} \right\} \mathrm{dt} = \\ [(\boldsymbol{p}(t_2) - \boldsymbol{p}_2) \cdot \delta \boldsymbol{q}(t_2) - (\boldsymbol{q}(t_2) - \boldsymbol{q}_2) \cdot \delta \boldsymbol{p}(t_2)] - \\ [(\boldsymbol{p}(t_1) - \boldsymbol{p}_1) \cdot \delta \boldsymbol{q}(t_1) - (\boldsymbol{q}(t_1) - \boldsymbol{q}_1) \cdot \delta \boldsymbol{p}(t_1)] \end{aligned} \tag{3.4}$$

which has no subsidiary conditions. Performing an integration by parts of the terms involving the time derivative explicitely, Eq.(3.4) may be simplified as:

$$\int_{t_1}^{t_2} \{ \delta (\mathcal{L} - \boldsymbol{p} \cdot \boldsymbol{v}) + \boldsymbol{f} \cdot \delta \boldsymbol{q} + \delta \dot{\boldsymbol{q}} \cdot \boldsymbol{p} - \delta \dot{\boldsymbol{p}} \cdot \boldsymbol{q} \} \, \mathrm{dt} = [\boldsymbol{p}_b \cdot \delta \boldsymbol{q} - \boldsymbol{q}_b \cdot \delta \boldsymbol{p}]_{t_1}^{t_2} \tag{3.5}$$

where $b = 1$ for $t = t_1$ and $b = 2$ for $t = t_2$. This result is analogous to the Hu-Washizu (general 3 field) form [8]. Each of the variables, $\boldsymbol{q}$, $\boldsymbol{v}$, and $\boldsymbol{p}$, appear as independent fields and have no continuity requirements at the time boundaries. The test functions $\delta\boldsymbol{p}$ and $\delta\boldsymbol{q}$, however must be differentiable on the interval and continuous at the boundary.

The Principle of Virtual Work is obtained from the three field form when the compatibility conditions and displacement boundary conditions are satisfied *a priori* . Moreover, the momentum field may be eliminated from the domain integral assuming the compatibility conditions are also satisfied for the test functions, i.e.: $\delta\boldsymbol{v} = \delta\dot{\boldsymbol{q}}$. In this way we obtain:

$$\int_{t_1}^{t_2} (\delta\mathcal{L} + \boldsymbol{f} \cdot \delta\boldsymbol{q})\mathrm{d}t = \boldsymbol{p}_b \cdot \delta\boldsymbol{q}|_{t_1}^{t_2} \tag{3.6}$$

This expression corresponds to Hamilton's principle in the configuration space and obviously has only one independent field, $\boldsymbol{q}$ in the domain. It is henceforth referred to as the primal or kinematic form of Hamilton's principle.

As with the locking phenomenon, which is well known in elasto-statics of volumetrically rigid media or shear rigid thin plates, typical of kinematic formulations, the kinematic form of Hamilton's principle is conditionally stable [5]. We can avoid the resulting restriction on the time step either through selective reduced integration or by utilizing a mixed formulation. In fact, if we choose to satisfy the constitutive relations *a priori*, we may eliminate the velocity as a function of the momentum, and a two field mixed form is obtained. This is accomplished through a Legendre transformation, introducing the Hamiltonian function, $H(\boldsymbol{p}, \boldsymbol{q}, t) = \boldsymbol{p} \cdot \boldsymbol{v} - \mathcal{L}(\boldsymbol{v}, \boldsymbol{q}, t)$, in which $\boldsymbol{v} = \boldsymbol{v}(\boldsymbol{p}, \boldsymbol{q}, t)$ is understood. Then the two field mixed form involving $\boldsymbol{p}$ and $\boldsymbol{q}$ may be written as:

$$\int_{t_1}^{t_2} \{-\delta H + \boldsymbol{f} \cdot \delta\boldsymbol{q} + \delta\dot{\boldsymbol{q}} \cdot \boldsymbol{p} - \delta\dot{\boldsymbol{p}} \cdot \boldsymbol{q}\}\,\mathrm{d}t = [\boldsymbol{p}_b \cdot \delta\boldsymbol{q} - \boldsymbol{q}_b \cdot \delta\boldsymbol{p}]_{t_1}^{t_2} \tag{3.7}$$

In this form $\boldsymbol{p}$ and $\boldsymbol{q}$ are treated as coordinates in the phase space, and are not required to be differentiable. It is important to note that since the test and the trial functions have different continuity requirements, the test functions $\delta\boldsymbol{p}$ and $\delta\boldsymbol{q}$ are properly understood to be weighting functions, independent of the variations of $\boldsymbol{p}$ and $\boldsymbol{q}$. The numerical implementation of this mixed form is unconditionally stable.

4 Weak Form of Constraints

For the purposes of illustration we consider here only the kinematic and two field mixed forms, and the class of constraints which can be expressed as:

$$\boldsymbol{\psi}(\dot{\boldsymbol{q}}, \boldsymbol{q}, t) = \mathbf{A}(\boldsymbol{q}, t) \cdot \dot{\boldsymbol{q}} + \boldsymbol{a}(\boldsymbol{q}, t) = 0 \tag{4.1}$$

These equations imply the following constraint on the virtual displacements:

$$\mathbf{A} \cdot \delta\boldsymbol{q} = 0 \tag{4.2}$$

Equation Eq.(4.1) can be either nonholomonic, or the time derivative of a holonomic constraint. In order to enforce Eq.(4.1) and Eq.(4.2), we cast them in weak form with a convenient choice of test functions. Let $\boldsymbol{\mu}$ be the Lagrange multipliers. We then weight Eq.(4.1) with the variation $\delta\boldsymbol{\mu}$ and Eq.(4.2) with the time derivative $\dot{\boldsymbol{\mu}}$, obtaining:

$$\int_{t_1}^{t_2} \left(\delta\boldsymbol{\mu} \cdot \boldsymbol{\psi} - \dot{\boldsymbol{\mu}} \cdot \frac{\partial \boldsymbol{\psi}}{\partial \dot{\boldsymbol{q}}} \cdot \delta \boldsymbol{q} \right) \mathrm{d}t = 0 \tag{4.3}$$

The benefit of this form is that it allows an integration by parts that reduces the continuity requirements for the Lagrangian multipliers. Combining Eq.(4.2) with the kinematic form Eq.(3.6) we obtain a modified form:

$$\int_{t_1}^{t_2} (\delta\bar{\mathcal{L}} + \delta\boldsymbol{q} \cdot \bar{\boldsymbol{f}})\mathrm{d}t = \delta\boldsymbol{q} \cdot \bar{\boldsymbol{p}}_b|_{t_1}^{t_2} \tag{4.4}$$

where:

$$\bar{\mathcal{L}} = \mathcal{L} + \boldsymbol{\mu} \cdot \boldsymbol{\psi} \qquad \bar{\boldsymbol{p}} = \boldsymbol{p} + \boldsymbol{\mu} \cdot \frac{\partial \boldsymbol{\psi}}{\partial \dot{\boldsymbol{q}}} \qquad \bar{\boldsymbol{f}} = \boldsymbol{f} + \boldsymbol{f}_c \tag{4.5}$$

and $\boldsymbol{f}_c = \boldsymbol{\mu} \cdot \left(\frac{\mathrm{d}}{\mathrm{d}t} \frac{\partial \boldsymbol{\psi}}{\partial \dot{\boldsymbol{q}}} - \frac{\partial \boldsymbol{\psi}}{\partial \boldsymbol{q}} \right)$. The Eq.(4.5) is the modified Hamilton's principle for the constrained systems and $\bar{\mathcal{L}}$, $\bar{\boldsymbol{p}}$, $\bar{\boldsymbol{f}}$ are respectively the modified Lagrangian function, the modified generalized momenta and the external forces modified by the reactions due to the nonholonomic constraints [9]. The constraint reactions $\boldsymbol{f}_c$ are just the weighted integrability conditions of the constraint equations.

It is interesting to note that $\bar{\boldsymbol{p}}$ are the generalized momenta associated with the modified Lagrangian, in fact it can easily be seen that $\bar{\boldsymbol{p}} = \frac{\partial \bar{\mathcal{L}}}{\partial \dot{\boldsymbol{q}}}$. We can therefore define the modified Hamiltonian function as $\bar{H} = \bar{\boldsymbol{p}} \cdot \dot{\boldsymbol{q}} - \bar{\mathcal{L}}$ and obtain a mixed form, following the same procedure as in the preceding section:

$$\int_{t_1}^{t_2} (-\delta\bar{H} + \delta\boldsymbol{q} \cdot \bar{\boldsymbol{f}} + \delta\dot{\boldsymbol{q}} \cdot \bar{\boldsymbol{p}} - \delta\dot{\bar{\boldsymbol{p}}} \cdot \boldsymbol{q})\mathrm{d}t = (\delta\boldsymbol{q} \cdot \bar{\boldsymbol{p}}_b - \delta\bar{\boldsymbol{p}} \cdot \boldsymbol{q}_b)|_{t_1}^{t_2} \tag{4.6}$$

Provided that velocity is eliminated from the expression of the reaction force, the modified momenta $\bar{\boldsymbol{p}}$, the generalized coordinates $\boldsymbol{q}$, and the multipliers $\boldsymbol{\mu}$, become the independent fields in this formulation. The true momenta may be recovered from the modified momenta by a projection.

5 Numerical Examples

In order to demonstrate the utility of the various methods discussed in the preceding sections, a few simple examples have been analyzed. All of these concern rigid body dynamics. Specifically, we consider the free tumbling body, the spinning top, as an example of a holonomically constrained system, and the rolling coin, as an example of nonholonomic constraints. Each of these cases is strongly nonlinear and constitutes a good test for the numerical methods. The coordinates in each example are the incremental position of the center of mass, and the incremental finite rotation of the body, with respect to a reference configuration. In this way, the singularity associated with the parametrization of finite rotation is avoided. A comparison of these solutions to those obtained by variable step Runge Kutta, showed a good behaviour of the methods [14].

Here, due to space limitation, we report only two numerical examples: the stability analysis of the vertical spinning top and the falling chain simulation.

The first example shows that the primal form is conditionally stable while unconditional stability is supplied by reduced order integrations or by mixed forms. This example gives a new interpretation of numerical instability as locking phenomenon typical of kinematics formulation as it is well recognized in elasto-statics.

For a top spinning at its critical speed, the eigenvalues of the system are purely imaginary. The exact eigenvalues of the system considered are $\pm i\frac{5}{3}$. Table 1 summarizes the numerical results obtained at the critical speed for exact integration. The influence of time step on the real part of the eigenvalues is clear. While the effect is not as strong in the higher order elements, the trend is the same.

The results for reduced order integration are presented in Table 2. It should be noted that the real parts of the eigenvalues are essentially zero, and are insensitive to the time step. For the case of a vertical spinning top or other simple system, it is straightforward to choose the degree of reduced integration required to follow the physics of the problem. However this is not the case in general.

For the mixed form, the stability boundary obtained numerically coincides with the physical boundary, without resorting to reduced order integration.

The problem of the falling chain, treated by Kamman and Huston [15], is shown in Figure 1. The chain consists of fifteen identical bars, connected by spherical joints. It is released at points B and C, and allowed to fall under the influence of gravity. The shape of the chain is calculated for times between 0 and 1 second. The input data for this problem are as follows:

$$\begin{aligned} \mathrm{M} &= .4565\,\mathrm{slugs} &&= \mathrm{Mass} \\ \mathrm{L} &= 1.0\mathrm{ft.} &&= \mathrm{Length} \\ \mathrm{I}_t &= .0308\mathrm{slug}-\mathrm{ft}^2 &&= \mathrm{Transverse\ Inertia} \\ \mathrm{I}_a &= .0001\mathrm{slug}-\mathrm{ft}^2 &&= \mathrm{Axial\ Inertia} \end{aligned} \tag{5.1}$$

Here, the axial inertia is simply set to some nonzero number for the planar problem. Figure 2, shows a plot of the vertical displacement of the center bar vs. time. This clearly shows, the reflection point at t=.47, and another point at t=.69, where the bar stops moving up and begins to fall again. The chain falls, as shown in Figure 3, and the center link reaches a low point at t=.47 second. At this point, the motion reflects and the center bar starts to move upward. The plot of the motion from t=.47 seconds until t=1 second is shown in Figure 4. These results are in complete agreement with those presented by Kamman and Huston [15].

6 Conclusions

Kinematic and mixed forms for dynamics are derived form a very general framework and are implemented using finite elements in the time domain. The new approach to deal with constrained systems is shown to be efficient and very general. The numerical examples indicate that the methods presented are quite promising.

Table 1 Eigenvalues for Vertical Top - Exact Integration

Eigenvalues for Exact Integration						
	Two Nodes		Three Nodes		Four Nodes	
Δt	Real	Imag	Real	Imag	Real	Imag
0.02	1.6031E-02	1.6663	2.7599E-04	1.6666	1.7240E-05	1.6666
0.04	3.2027E-02	1.6654	1.1032E-03	1.6666	4.1751E-05	1.6666
0.06	4.7952E-02	1.6638	2.4795E-03	1.6666	1.4113E-04	1.6666
0.08	6.3772E-02	1.6617	4.4010E-03	1.6666	3.3443E-04	1.6666
0.10	7.9453E-02	1.6590	6.8623E-03	1.6665	6.5150E-04	1.6666
0.12	9.4963E-02	1.6557	9.8564E-03	1.6664	1.1222E-03	1.6666
0.14	0.1102	1.6518	1.3374E-02	1.6663	1.7753E-03	1.6666
0.16	0.1253	1.6474	1.7404E-02	1.6661	2.6385E-03	1.6666
0.18	0.1401	1.6424	2.1934E-02	1.6658	3.7381E-03	1.6666
0.20	0.1546	1.6370	2.6949E-02	1.6653	5.0991E-03	1.6666

Table 2 Eigenvalues for Vertical Top - Under Integration

Eigenvalues for Reduced Integration						
	Two Nodes		Three Nodes		Four Nodes	
Δt	Real	Imag	Real	Imag	Real	Imag
0.02	2.3881E-05	1.6665	2.3261E-05	1.6666	1.6713E-05	1.6666
0.04	2.3211E-05	1.6660	1.0527E-05	1.6666	1.3681E-05	1.6666
0.06	1.6634E-05	1.6652	1.3681E-05	1.6666	7.9443E-06	1.6666
0.08	1.0443E-05	1.6642	7.7153E-06	1.6666	8.0284E-06	1.6666
0.10	1.2153E-05	1.6628	8.0605E-06	1.6666	7.6044E-06	1.6666
0.12	1.3460E-05	1.6611	8.0266E-06	1.6666	7.4198E-06	1.6666
0.14	7.2016E-06	1.6591	7.7597E-06	1.6665	8.3770E-06	1.6666
0.16	7.5157E-06	1.6568	7.4664E-06	1.6664	6.2104E-06	1.6666
0.18	7.6967E-06	1.6543	7.4116E-06	1.6663	6.1258E-06	1.6666
0.20	7.7657E-06	1.6514	7.8648E-06	1.6662	6.0095E-06	1.6666

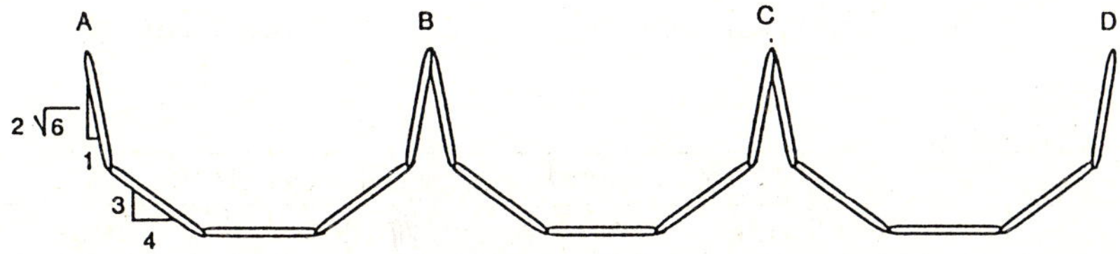

Figure 1 Fifteen Link Chain

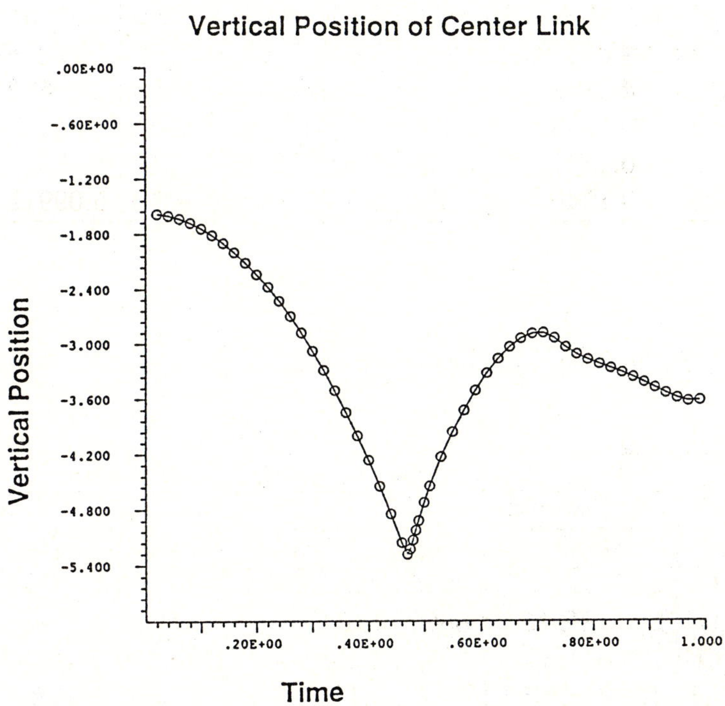

Figure 2 Vertical Displacement of Center Link

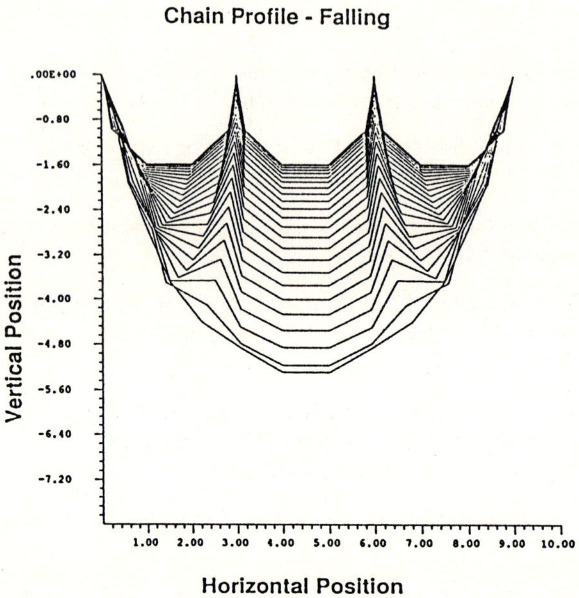

Figure 3 Chain Profile - Falling

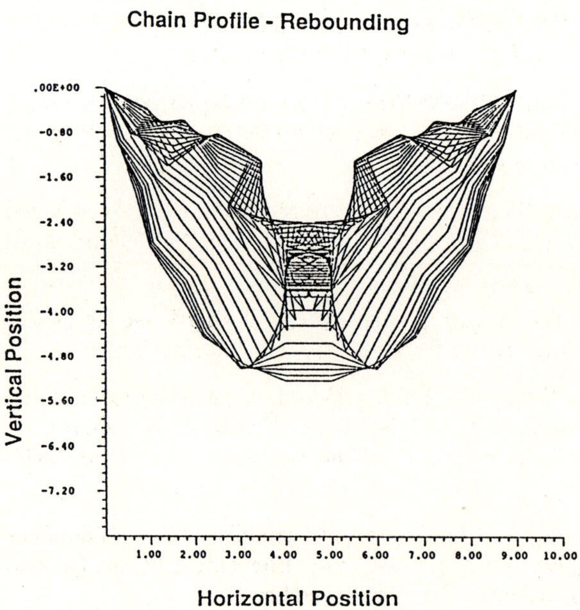

Figure 4 Chain Profile - Rebounding

7 References

1. C.D. Bailey, "Application of Hamilton's Law to Forced, Damped, Stationary Systems", *Journal of Sound and Vibration*, Vol. 75, pp. 117-126, 1981.

2. T.E. Simkins, "Finite Element for Initial Value Problems in Dynamics", *AIAA Journal*, Vol. 13, pp. 1154-1157, 1981.

3. M. Baruch and R. Riff, "Hamilton's Principle, Hamilton's Law, 6^n correct formulations", *AIAA Journal*, Vol. 21, pp. 687-692, 1982.

4. M. Borri, M. Lanz and P. Mantegazza, "Helicopter Rotor Dynamics by Finite Element Time Discretization", *L'Aerotecnica Missili e Spazio*, Vol. 60, pp. 193-200, 1981.

5. M. Borri, G.L. Ghiringhelli, M. Lanz, P. Mantegazza and T. Merlini, "Dynamic Response of Mechanical Systems by a Weak Hamilton's Formulation", *Computers & Structures*, Vol. 20, pp. 495-508, 1985.

6. T. Levi Civita and V. Amaldi, Lezioni di Meccanica Razionale, Zanichelli, Bologna 1979.

7. C. Lanczos, "The Variational Principles of Mechanics", University of Toronto, Toronto 1964.

8. K. Washizu, "Variational Methods in Elasticity and Plasticity", third edition, Pergamon 1980.

9. Iu. I. Neimark and N.A. Fufaev, "Dynamics of Nonholonomic System", American Mathematical Society, Providence, Rhode Island 1972.

10. M. Borri and P. Mantegazza, "Finite Time Elements Approximation of Dynamics of Nonholonomic Systems", paper presented at AMSE Congress, Williamsburg, Virginia, August 1986.

11. M. Borri and S.N. Atluri, "Time Finite Element Method for the Constrained Dynamics of a Rigid Body", Proceedings of ICES-88, Atlanta, Georgia, April 1988.

12. M. Iura, M. Borri, and S.N. Atluri, "Analysis of Traveling Wave Responses of Structures", Proceedings of ICES-88, Atlanta, Georgia, April 1988.

13. P.E. Nikravesh and E.J. Haug, "Generalized Coordinate Partioning for Analysis of Mechanical Systems with Nonholonomic Constraints", *Journal of Mechanisms, Transmissions, and Automation in Design*, September 1983, Vol. 105, pp. 379-384.

14. F. J. Mello, "Weak Formulations in Analytical Dynamics, with Applications to Multi-Rigid-Body Systems, Using Time Finite Elements", Thesis, Georgia Institute of Technology, Atlanta, Georgia, December 1989.

15. Kamman, J.W. and Huston, R.L., Dynamics of Constrained Multibody Systems, *Journal of Applied Mechanics*, Vol. 51, pp. 899-903, 1984.

On Static Stability of Elastic Nonconservative Systems with Small Damping

A. S. Bratus

Department of Applied Mathematics,
Moscow Railway Engineering Institute

Summary

Nonconservative systems with finite and infinite numbers of degrees of freedom are considered . Coefficients of the equation that describe the mechanical behavior of the systems depend on loading parameter. Besides such systems augmented perturbed system taking into account additional terms of small viscous damping. It is well known that addition of dissipative or gyroscopic forces to conservative system does not violate its stability. In general this result does not follow for nonconservative systems. The set of viscous damping forces that does not violate the stability of nonconservative system are described in form of necessary and sufficient conditions. Account have done for the case of finite numbers of degrees of freedom, but all results was kept for infinite numbers.

Statement of the problem

Consider the second order equation is used to describe the mechanical behavior of systems, taking into account additional terms of small viscous damping

$$A(p)u(t) + \varepsilon B(p)\frac{du}{dt} + \frac{d^2u}{dt^2} = 0, \qquad t > 0 \tag{1}$$

where t is time, $u(t)$ is the state vector function of order $n \geq 2$, $A(p)$, $B(p)$ are real matrices depends

N. V. Banichuk · D. M. Klimov · W. Schiehlen (Eds.)
Dynamical Problems of Rigid-Elastic Systems and Structures
IUTAM Symposium Moscow, USSR, May 23-27, 1990

analytically on real loading parameter p , ε is a small parameter having a sense of intensity of damping. The matrix $B(p)$ determine structure of damping forces.

If the viscous damping forces disappear, the equation (1) has a form

$$A(p)u(t) + \frac{d^2u}{dt^2} = 0, \qquad t > 0 \tag{2}$$

Ordinary in mechanics the matrix $A(p)$ has a presentation $A(p) = S + pN$, where S, N are symmetrical matrixes with constant coefficients, i.e. loading parameter include by linear manner. If S is positive definite and N is negative definite matrices, then addition to system (2) dissipative or gyroscopic forces does not violate its stability. This is well known result called as Kelvin - Tait- Chetayev theorem [1]. But this is not truth for nonconservatine systems. Here the small damping can deduce to destabilizing effects [2 - 6].

Eigenvalue problems

For investigation stability of the system (2) the special solution of this system is sought in the form

$$u(t) = ve^{i\omega t} \tag{3}$$

where v is the amplitude vector, ω is the frequency, i is the imaginary unit. Putting the $u(t)$ into (2) the following eigenvalue problem depends on loading parameter p are obtained

$$A(p)v = \omega^2(p)v \tag{4}$$

The following basic assumptions are introduced.

(1) There exists a value p_0 of loading parameter p, such that for $p < p_0$ all eigenvalues $\omega_j^2(p)$, $(j = 1, 2, \ldots, m)$ the problem (4) are real.

(2) For $p = p_0$ the matrix $A(p_0)$ has a simple eigenvalue $\omega_i^2(p_0) = 0$ and for $p > p_0$ these eigenvalue became complex so that $\mathrm{Im}\omega_1(p) < 0$.

(3) The other ω_j^2, $(j = 2, 3, \ldots, n)$ remain real in sufficiently small neighborhood of p_0.

From assumption 1-3 follows that for $p = p_0$ takes a place the bifurcation and its has a static (divergence) character.

Substituting (4) into perturbed system (1) we get

$$A(p)v + i\varepsilon\omega B(p)v = \omega^2 v \tag{5}$$

The ideal and nonideal perturbations

The value p_ε is termed the critical value of the loading parameter for system (1), if it is the lowest bound of the values p for which there exists at least one frequency ω of (5) with negative imaginary part. The set of points $(\varepsilon, p_\varepsilon)$ in the $\varepsilon - p$ plane defines a curve that separates the stable domain from instable domain. It is called the critical curve.

Denote by p_d the limit of p_ε when $\varepsilon \to +0$. It is known [2 - 6] that any cases $p_d \le p_0$. If $p_d = p_0$ the corresponding perturbations of the system (2) in form (1) will be called ideal. In opposite case will be called the perturbations nonideal (defective). The necessary conditions of ideal perturbation have been found in [7] for the case of dynamic bifurcation (flutter). Our purpose to find out anologous condition for the case of static bifurcation (divergence).

The necessary and sufficient condition for ideal perturbation

Together with the eigenvalue problem (4) consider the ajoint eigenvalue problem

$$A^T(p)z = \omega^2 z \tag{6}$$

where $A^T(p)$ is adjoint matrix for $A(p)$.

The following assumption takes a place.

Theorem. Let for $p = p_0$ the simple eigenvalue $\omega_1^2(p_0)$ of problem (4) is equal to zero and the conditions 1 - 3 are satisfied. Then for ideal perturbation system (2) in form (1) necessary and sufficient that the following conditions are held

$$(B(p_0)v_i^0,\ z_i^0) \geq 0, \qquad i = 1, 2, \dots, n \tag{7}$$

$$(A_p(p_0)v_1^0,\ z_1^0) < 0$$

With that for sufficiently small $\varepsilon > 0$ the critical curve in $\varepsilon - p$ plane has an expansion

$$p_\varepsilon = p_0 + 0(\varepsilon^4)$$

Here the bracket denote the scalar product, v_i^0, z_i^0, (i = = 1, 2, ..., n) are eigenvectors of direct and adjoint eigenvalue problems (4) and (6) correspondently for $p = p_0$,

$$A_p(p_0) = \left(\frac{d}{dp} A(p)\right)_{p=p_0}$$

Corollary 1. If the conditions (7) does not valid then for any sufficiently small $\varepsilon > 0$ the perturbed system (1) will be unstable on $p = p_0$.

Corollary 2. Let the conditions (7) are valid in strong form then the perturbed system (1) will be stable for sufficiently small ε and any $p < p_0$.

The last proposition represent the generalization on Kelvin--Tait-Chetayev theorem for nonconservative case.

<u>Example</u>

Let the motion of the mechanical system have described by equations

$$\ddot{u}_1 = ku_1 - pu_1 + pu_2 + \varepsilon(\dot{u}_1 b_{11} + \dot{u}_2 b_{22})$$

$$\ddot{u}_2 = 16ku_2 - 4pu_2 + \varepsilon(\dot{u}_1 b_{21} + \dot{u}_2 b_{22})$$

where p is a loading parameter, $k > 0$, $\varepsilon > 0$, b_{ij} are real $i, j = 1, 2$.

The corresponding eigenvalue problem (4) have the solutions $\omega_1^2 = k - p$, $\omega_2^2 = 4k - p$. If $3b_{11} - b_{21} > 0$, $3b_{22} + b_{21} > 0$ and consequently $b_{11} + b_{22} > 0$, then all assumptions of Corollary 2 are fulfilled for $p_0 = k$ and $v_1^0 = (1, 0)^T$, $v_2^0 = (1, 3)^T$, $z_1^0 = (3, -1)^T$, $z_2^0 = (0, 1)^T$. This result can be immediately verify since the characteristic equation has a form

$$\omega^4 - i\varepsilon\omega^3(b_{11}+b_{22}) - \omega^2(5k - 2p + \varepsilon^2(b_{11}b_{22}-b_{12}b_{21})) +$$

$$+ i\varepsilon\omega(b_{11}(4k-p) + b_{22}(k-p) - b_{21}p) + (4k-p)(k-p) = 0$$

After application of Routh-Hurwitz criterion for $p_0 = k$ we get for sufficiently small ε

$$b_{11} + b_{22} > 0, \quad 3b_{11} - b_{21} > 0, \quad 3b_{22} + b_{21} + 0(\varepsilon^2) > 0$$

Finally notice that the part of presented results was obtained together with prof. N.V.Banichuk.

References

1. Chetayev, N.G.: Stability of Motion, Nauka, Moscow 1965. English translation published by Pergamon Press, Oxford, London, New York, 1961.

2. Ziegler, H.: Stabilitätskriterien der Elastomechanik, Ing. Arch. 20 (1952) 49-56.

3. Bolotin, V.V.: Nonconservative Problems of the Theory of Elastic Stability, Fizmatgiz, Moscow, 1961. English translation published by Pergamon Press, New York, 1962.

4. Hermann, G.; Jong, I.C.: On the destabilizing effect of damping in nonconservative elastic systems. J. Appl. Mech. Trans. ASME 32 (1965) 592-597.

5. Andeichikov, I.P.; Yudovich, V.I.: On stability of viscoelastic rods, Mech. Solids 2 (1974) 78-87, (in Russian).

6. Denisov, G.G.; Novikov, V.V.: On stability of elastic systems with small internal damping, Mech. Solids, 3 (1978) 41-47 (in Russian).

7. Banichuk, N.V.; Bratus, A.S.; Myshkis, A.D.: Dynamic stability of nonconservative mechanical systems with small damping, Mech. Struct. Mach. 18 (3) (1990) 81-95.

Control of Elastic Robots

H. Bremer, F. Pfeiffer

Institute B of Mechanics, Technical University of Munich, FRG

Summary

A robot system is considered which consists of two elastic links (bending and torsional deflection) and three fexible joints. This leads to six degrees of freedom (dof) for the rigid body motion of the system, three of which are assigned to the motor torque input. The remaining three dof.describe the gross motion of the system in space. They are superimposed by small elastic deflections which are calculated using a Ritz series approximation. The differential equations of the interconnected rigid body and elastic motion are highly nonlinear. The aim of the present investigation is to evaluate an optimal endpoint control (gripper movement) for a prescribed path in space.

1 Introduction

In order to increase the payload to mass ratio of an industrial robot, the elasticity not only of the joints but also of the links has to be considered from the very first step in the conceptional phase, [3], [7], [8], [9], [18], [26], [31]. Once the elastic behaviour is known, the total equipment may be used as a force measurement device itself, yielding a simple force control device. The main element to be considered is therefore the elastic body. Here, a Ritz series approximation will enter the description which makes use of the *eigenfunctions* of the separated links, [5]. It does not matter where these eigenfunctions come from, e.g. by using finite element methods or by measurements. Thus, an Euler-Bernoulli-Beam will be a good representative for mainly beamlike elastic bodies in the subsequent discussion, [11], [12], [20], [21].

First experiments were performed on a one-beam-robot about ten years ago at the Munich Institute, [33], which already showed some basic properties of elastic mechanical systems. The one-element-robot consisted of a one meter long rectangular shaped aluminium beam (1kg) with a heavy payload (10kg). The frequency response showed a strong coupling between torsional and bending deflection due to geometrical nonlinearities, which in the present are often discussed as "dynamical stiffening" effects, [19], and vanishing amplitudes of the shoulder at certain frequencies along with a peak in the bending deflection: In this case the shoulder remains at rest while the total input energy is used to exite bending as if the beam were clamped at the shoulder. This effect is well known in

N.V. Banichuk · D. M. Klimov · W. Schiehlen (Eds.)
Dynamical Problems of Rigid-Elastic
Systems and Structures
IUTAM Symposium Moscow, USSR, May 23-27, 1990

rotor dynamics, for instance, yielding instability in some cases. One may therefore state that a lot of additional physical phenomena occur in elastic systems when compared to rigid body systems, and special attention must be given to mathematical modelling.

2 Plant

There is of course a long way from simple one beam considerations to a more realistic system as it is shown in Fig. 2.1. It depicts the Munich laboratory robot, consisting of

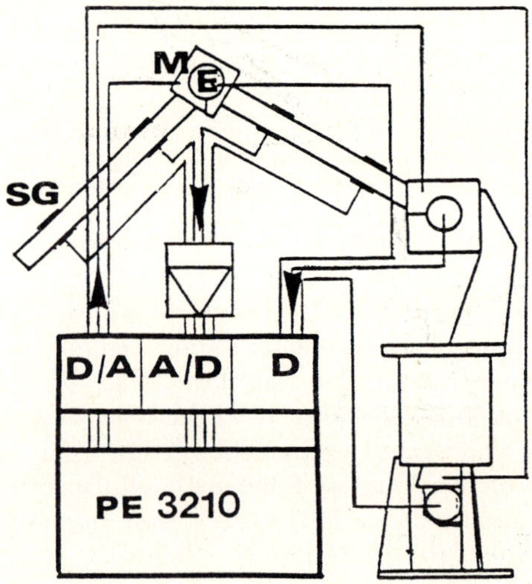

Fig. 2.1: The Munich Elastic Robot

- two flexible arms,
- connected by shoulder and ellbow,
- thus having six degrees of freedom of rigid body motion, two at the shoulder and one for the elbow (both motor-input/gear-output)
- and an infinite number of degrees of freedom for bending and torsional deflections.
- Bending and torsion are measured by strain gauges, motor angles by incremental devices.
- Measurements are then used to calculate control torques by a digital computer
- which are fed back to the joint motors as analog signals.

In order to calculate the corresponding equations of motion a most effective procedure is used, [4], [5], [14], which projects the momentum $\boldsymbol{p}$ and angular momentum $\boldsymbol{L}$ derivations of infinitesimal element dm into the unconstrained directions by means of corresponding

Jacobians. Here, $\boldsymbol{v}$ and $\boldsymbol{\omega}$ are the 3-dimensional absolute velocities, given in a reference frame which rotates with $\overline{\omega}$, and $\boldsymbol{q}$ contains the generalized coordinates:

$$\sum_{i=1}^{n} \int_{K_i} \left\{ \begin{pmatrix} \dfrac{\partial \boldsymbol{v}_i}{\partial \dot{\boldsymbol{q}}} \\ \dfrac{\partial \boldsymbol{\omega}_i}{\partial \dot{\boldsymbol{q}}} \end{pmatrix}^T \begin{pmatrix} (d\dot{\boldsymbol{p}} + \tilde{\overline{\omega}} d\boldsymbol{p} - d\boldsymbol{f}^e)_i \\ (d\dot{\boldsymbol{L}} + \tilde{\overline{\omega}} d\boldsymbol{L} - d\boldsymbol{l}^e)_i \end{pmatrix} \right\} = o \tag{2.1}$$

The mathematical model is calculated using rigid body coordinates on the one hand side and distributed coordinates on the other hand.

The rigid body coordinates are the motor angle γ, the gear output angle α (both connected by the gear transmission ratio $(\partial\gamma/\partial\alpha)$) and the arm angle β. The joint elasticity has been measured separately by special devices. It connects the motor-gear-output angle α and arm angle β for every motor-gear-unit.

For the elastic coordinates a Ritz expansion is used. Here, the bold faced symbols represent admissible shape functions and the corresponding time variables are used as generalized coordinates. Note that if z is the long axis of the beam, the second order elongation $\overline{w}$ enters the description even in the case that the beam is inextensible.

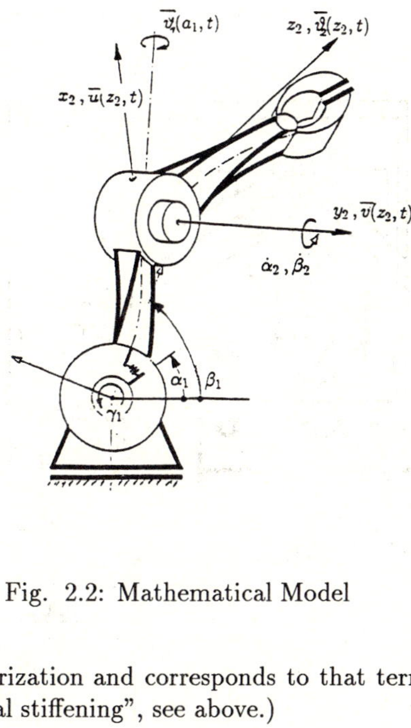

Fig. 2.2: Mathematical Model

(This is merely the result of a correct linearization and corresponds to that term which nowadays is broadly discussed as "dynamical stiffening", see above.)

$$\begin{pmatrix} \alpha \\ \beta \\ \gamma \\ \hline \overline{u}(z,t) \\ \overline{v}(z,t) \\ \overline{w}(z,t) \\ \overline{\vartheta}(z,t) \end{pmatrix} = \left[\begin{array}{cc|ccc} 1 & 0 & 0 & 0 & 0 \\ 0 & 1 & 0 & 0 & 0 \\ \left(\dfrac{\partial\gamma}{\partial\alpha}\right) & 0 & 0 & 0 & 0 \\ \hline 0 & 0 & \boldsymbol{u}(z)^T & 0 & 0 \\ 0 & 0 & 0 & \boldsymbol{v}(z)^T & 0 \\ 0 & 0 & \left(\dfrac{\partial\overline{w}}{\partial\boldsymbol{y}_u}\right) & \left(\dfrac{\partial\overline{w}}{\partial\boldsymbol{y}_v}\right) & 0 \\ 0 & 0 & 0 & 0 & \boldsymbol{\vartheta}(z)^T \end{array} \right] \boldsymbol{q}, \tag{2.2}$$

$$\overline{w} = -(\boldsymbol{y}_u^T \; \boldsymbol{y}_v^T)\frac{1}{2}\int\left[\left(\frac{du}{dz}\right)^2 + \left(\frac{dv}{dz}\right)^2\right]dz\begin{pmatrix}\boldsymbol{y}_u\\ \boldsymbol{y}_v\end{pmatrix} \tag{2.3}$$

where $\boldsymbol{q}^T = (\alpha\,\beta\,\boldsymbol{y}_u^T\,\boldsymbol{y}_v^T\,\boldsymbol{y}_\vartheta^T)$ for every motor-gear-link-unit.

3 Control

3.1 Open Loop Control

The outlined procedure yields the equations in common form, the vector $\boldsymbol{g}$ containing all nonlinearities due to coriolis and centifugal forces as well as the control input $\boldsymbol{u}$,

$$\boldsymbol{M}(\boldsymbol{q})\ddot{\boldsymbol{q}} + \boldsymbol{g}(\boldsymbol{q}, \dot{\boldsymbol{q}}, \boldsymbol{u}) = 0, \tag{3.1}$$

[3], [6], [16], [17], [24], [25]. It is obviously not possible to calculate a general nonlinear control concept for the robot for any arbitrary path. Thus, the control design is split up into three stages, the first of which is the generation of an open loop control for a prescribed path, [13], [10], [27]. For this purpose only the rigid body coordinates are retained. The prescribed path and orientation are gathered within a vector $\boldsymbol{x}$, which is of course a function of the generalized coordinates $\boldsymbol{q}_{RB}$ and forms a closed kinematical loop – starting at an inertial reference point, going around the joints and links and closing the loop with the prescribed path vector, [2], [15], [34].

3.1.1 Optimization

The calculation of corresponding open loop joint torques can easily be performed using the path parameter s as new variable, [18], [28]. The equation of motion then yields a first order differential equation with respect to the path variable, each component of which represents a constraint equation for the corresponding optimization problem. Additional constraints may be given if the joint angular velocities $\dot{\boldsymbol{q}}_{max}$ and/or the path velocity v_{max} itself are limited. Then the optimal torques may be calculated for every prescribed path using dynamical programming:

$$\int_{o}^{\sigma_e} f(z, z', s)ds \rightarrow min. \tag{3.2}$$

$$-\mid u_{i,max} \mid \leq A_i z' + B_i z + C_i \leq + \mid u_{i,max} \mid \tag{3.3}$$

$$z \leq min \left\{ \begin{pmatrix} \frac{\dot{q}^2_{max}}{q'^2_i} \\ \\ \frac{v^2_{max}}{\mid r' \mid^2} \end{pmatrix} \right. , \tag{3.4}$$

where $z = (\dot{s}^2/2)$, $(\ )' = \frac{\partial}{\partial z}(\ )$ and

$$A_i = \frac{1}{2} \boldsymbol{M}_i \boldsymbol{q}', \tag{3.5}$$

$$B_i = \boldsymbol{M}_i \boldsymbol{q}'' + \boldsymbol{q}'^T \left[\frac{\partial \boldsymbol{M}_i}{\partial \boldsymbol{q}}^T - \frac{1}{2} \frac{\partial \boldsymbol{M}}{\partial \boldsymbol{q}_i} \right] \boldsymbol{q}', \tag{3.6}$$

$$C_i = G_i, \tag{3.7}$$

G_i being the gravitational influence on the i-th coordinate, while $\boldsymbol{M}_i$ represents the i-th row of the Mass matrix $\boldsymbol{M}$.

The first contraint equation, eq.(3.3), yields a pair of parallels (minimum and maximum input) in the $z' - z$–plane for every value of the path parameter s. These pairs of parallels form a polygon which may additionally be bounded to the right by the second constraint equation, eq.(3.4). Using the path parameter s as third variable we are left with a 3-dimensional regular plane. The time optimal solution is found in the $z - s$–phase plane as the highest possible trajectory, i.e. starting from the left with maximum slope and reaching the endpoint with minimum slope, where in between the solution corresponds to the constraint of maximum possible joint angular velocity or path velocity, see also [1]. Once this phase curve is calculated, the corresponding joint torques are immediately obtained as function of time for open loop control.

3.1.2 Off Line Correction

Because the considered robot is extremely deformable, however, the above calculated open loop control will never be able to get the endpoint into the desired position. A certain correction torque will therefore be necessary, [13], [21]. Linearizing the equations with respect to the nominal path, an error $\boldsymbol{e}$ will arise as a nonhomogeneous term which includes nominal accelerations as well as gravitational effects:

$$\boldsymbol{q} = \begin{pmatrix} \boldsymbol{q}_{RB} \\ \boldsymbol{q}_{el} \end{pmatrix} = \begin{pmatrix} \boldsymbol{q}_{RB,N} \\ 0 \end{pmatrix} + \begin{pmatrix} \boldsymbol{y}_1 \\ \boldsymbol{y}_2 \end{pmatrix} = \boldsymbol{q}_N + \boldsymbol{y} : \tag{3.8}$$

$$\Rightarrow \boldsymbol{M}\ddot{\boldsymbol{y}} + \boldsymbol{P}\dot{\boldsymbol{y}} + \boldsymbol{Q}\boldsymbol{y} = \boldsymbol{e} + \boldsymbol{u}, \tag{3.9}$$

where N: "nominal input", RB: "Rigid Body", el: "Elastic". The endpoint error could then be calculated by means of a Jacobian which connects the prescribed path and its deviation: $\Delta\boldsymbol{x} = (\partial\boldsymbol{x}/\partial\boldsymbol{y})\boldsymbol{y}$. However, this procedure requires the solution of the non-homogenous differential equation (3.9). In a simpler approach it is assumed that neither deviation acceleration nor velocity will affect the endpoint error in a significant way, yielding

$$\boldsymbol{Q}\boldsymbol{y} = \boldsymbol{e} + \boldsymbol{u} \tag{3.10}$$

We are thus left with a simple procedure to calculate a correction input $\boldsymbol{u}$, taking into account the error of the arm coordinates and the elastic deflections. This assumption will be proved by experiments.

3.2 Closed Loop Control

In the third step, a superimposed input is calculated which only takes the deformations into account. Such a feedback control is of course to be calculated with known procedures such as Riccati-Optimization, yielding an optimal PD control, [5]. It should be mentioned, however, that due to gravitation (and nominal torques) the bending curvature along the trajectory can never be zero. It would therefore be contradictory to require a zero curvature along the optimal path. Thus, the closed control loop has to minimize the difference between the actual and the required curvature which once more is obtained from the correction model as an off-line calculation by eq. (3.10).

3.3 Force Control

Up to this point a free motion has been considered. If the robot has contact with a surface, however, the problem of force control will arise, [22], [23], [28], [29], [30]. This means that within the linearized equations certain constraint forces (or torques) have to be taken into account. This problem may be solved as follows: The Insertion of the desired nominal force into the open loop control calculation, i.e.

$$C_i = G_i - \boldsymbol{J}_i \boldsymbol{F}_N, \tag{3.11}$$

instead of eq.(3.5) ($\boldsymbol{J}_i$ being i-th row the Jacobian assigned to the nominal force $\boldsymbol{F}_N$), already fulfills the corresponding constraints $\boldsymbol{\Phi}(\boldsymbol{q}_N) = 0$ within the open loop calculation. Linearization then yields an additional constraint which is linear with respect to the deviation variable and which, using the chain rule of differentiation, may be interpreted as a deviation in the normal direction of the constraint plane:

$$\boldsymbol{\Phi}(\boldsymbol{y}) = \underbrace{\boldsymbol{\Phi}(\boldsymbol{q}_N)}_{0} + \left(\frac{\partial \boldsymbol{\Phi}}{\partial \boldsymbol{q}}\right)_N \boldsymbol{y} = \left(\frac{\partial \boldsymbol{\Phi}}{\partial \boldsymbol{x}}\right) \left(\frac{\partial \boldsymbol{x}}{\partial \boldsymbol{q}}\right)_N \boldsymbol{y} = \boldsymbol{n}^T \boldsymbol{J}\, \boldsymbol{y}. \tag{3.12}$$

The constraint force is then orthogonal to the constraint plane and is projected into the directions of possible motion by the corresponding Jacobian:

$$\boldsymbol{M}\ddot{\boldsymbol{y}} + \boldsymbol{P}\dot{\boldsymbol{y}} + \boldsymbol{Q}\boldsymbol{y} = \boldsymbol{u} - \left(\frac{\partial \boldsymbol{\Phi}}{\partial \boldsymbol{y}}\right)^T \lambda = \boldsymbol{u} - \boldsymbol{J}^T \underbrace{\boldsymbol{n}\lambda}_{\boldsymbol{f}} \ . \tag{3.13}$$

The control aim is now to cancel $\boldsymbol{f}$ because the desired force $\boldsymbol{F}_N : \boldsymbol{F} = \boldsymbol{F}_N + \boldsymbol{f}$ has already been attained by the nominal open loop calculation.

Combining (3.12) and (3.13) yields

$$\begin{pmatrix} \boldsymbol{M} & \boldsymbol{J}^T \\ \boldsymbol{n}^T\boldsymbol{J} & 0 \end{pmatrix} \begin{pmatrix} \ddot{\boldsymbol{y}} \\ \boldsymbol{f} \end{pmatrix} = \begin{pmatrix} -\boldsymbol{P}\dot{\boldsymbol{y}} - \boldsymbol{Q}\boldsymbol{y} \\ -2\frac{d}{dt}\left[\boldsymbol{n}^T\boldsymbol{J}\right] \dot{\boldsymbol{y}} + \frac{d^2}{dt^2}\left[\boldsymbol{n}^T\boldsymbol{J}\right] \boldsymbol{y} \end{pmatrix} + \begin{pmatrix} \boldsymbol{u} \\ 0 \end{pmatrix} \tag{3.14}$$

where the lower part follows from the second time derivative of the constraint equation.

4 Experimental Results

The outlined procedure is now proved by a comparison to simulation results in a simple experiment. The robot draws a straight line on a plane while the corresponding prescribed normal force increases. Normal force and friction between endpoint and surface are plotted in Fig. 4.1. The solid line corresponds to measurement and the dashed line corresponds to simulation results. (1: Normal force, 2: Friction force).

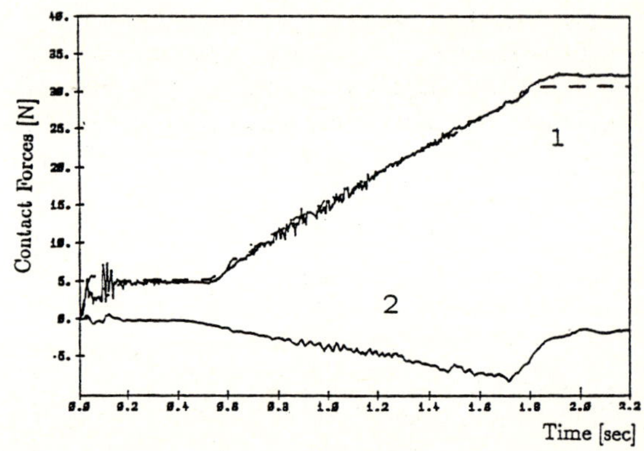

Fig. 4.1: Contact Forces

Due to increasing nominal force, the vertical deformation of both the upper arm and the forearm increases, too, reaching a constant value at the end of trajectory as can be seen in Fig. 4.2. In between, mainly the upper arm deforms in horizontal direction along with some torsional deflection. The deviation at the end of the considered maneuver is due to stiction. (1: Forearm vertical bending, 2: Upper arm torsion, 3: Upper arm horizontal bending, 4: Upper arm vertical bending).

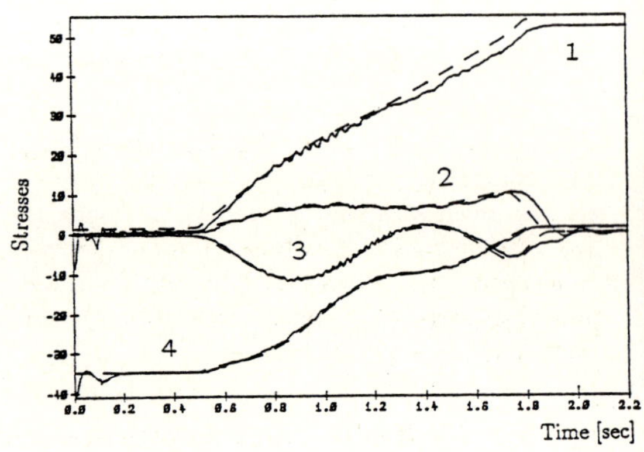

Fig. 4.2: Stresses

5 Conclusions

Much attention must be given to correct modelling, yielding a highly nonlinear equation of motion. Control is then calculated in three stages:

(1) If a path is prescribed, either as a free trajectory including gripper forces and torques (or not), or as a path along a given surface, an optimized open loop control is calculated off-line, yielding corresponding open-loop joint torques. To this purpose, the nominal state is calculated from given reference trajecory $\boldsymbol{r}_o$ via rigid body kinematics (RBK).

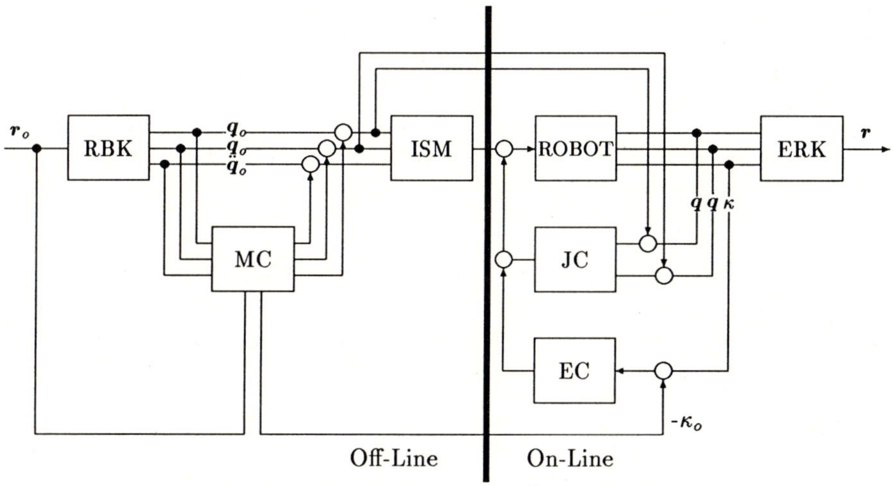

Fig. 5.1: Three Stage Control Concept

(2) The corresponding torques will of course not fulfill the requested task if elasticity is present. The deviation $\boldsymbol{y}$ of the nominal path is therefore to be compensated by a superimposed open-loop correction input which is obtained by a correction model (MC) and the inverse system model (ISM).

(3) One is finally left with the elastic deflection oscillations which are controlled in the last step (EC) using well known procedures. This is the only closed loop part, except a superimposed linear joint control (JC) which is necessary for rigid body stabilization. The controlled trajectory $\boldsymbol{r}$ is finally calculated using the total elastic robot kinematics (ERK).

References

[1] J.E. Bobrow, S. Dubowsky, J.S. Gibson: **Time-optimal Control of Robotic Manipulators Along Specified Paths.** Int. J. Robot. Res., Vol. 4, 3-17, 1985

[2] H. Brandl, R. Johanni, M. Otter: **An Algorithm for the Simulation of Multibody Systems with Kinematic Loops**, Proc. IFToMM Symp, Sevilla,1987.

[3] H. Bremer: **On the Dynamics of Flexible Manipulators.** Proc. 2nd IEEE Conf. Robotics and Automation, Rayleigh/USA, 1556 - 1560, 1987

[4] H. Bremer: **Über eine Zentralgleichung in der Dynamik**, ZAMM 68, 307 - 311, 1988

[5] H. Bremer: **Dynamik und Regelung mechanischer Systeme**, Teubner, Stuttgart, 1988.

[6] G. Demaria, B. Siciliano: **A Multilayer Approach to Control of a Flexible Arm.** Proc. 2nd IEEE Conf. Robotics and Automation, Rayleigh/USA, 774 - 778, 1987

[7] S. Dubowsky, J. Maatuk: **The Dynamic Analysis of Elastic Spatial Mechanisms.** J. Mech. Eng., 927 - 932, 1975

[8] S. Dubowsky, Z. Shiller: **Optimal Trajectories for Robotic Manipulators.** Proc. 5th CISM-IFToMM Symp. on Theory and Practice of Robots and Manipulators, 1985

[9] T. Fukuda: **Flexibility Control of Elastic Robot Arms.** J. Robotic Syst., 73 - 88, 1985

[10] E. Freund, H. Hoyer: **Das Prinzip nichtlinearer Systementkopplung mit Anwendung auf Industrieroboter.** Regelungstechnik 18, 80 - 87 und 116 - 126, 1980

[11] B. Gebler: **Modeling and Control of a Leightweight Robot**, Proc. 2nd Europ. Space Mech. & Tribology Symp., ESA-SP-231, 59-64, 1985

[12] B. Gebler: **Steuerung und Regelung für elastische Industrieroboter**, Fortschrittberichte VDI, Reihe 11, Nr. 98, VDI-Verlag, Düsseldorf 1987.

[13] B. Gebler, F. Pfeiffer: **A Multistage Approach to the Dynamics and Control of Elastic Robots**, Proc. of 1988 IEEE Conf. on Robotics and Automation, Philadelphia, PA.

[14] G. Hamel: **Theoretische Mechanik**, Springer (1967)

[15] M. Hiller, A. Kecskemethy: **A Computer-Oriented Approach for the Automatic Generation and Solution of the Equations of Motion for Complex Mechanisms**, Proc. 7th World Congr. Th. Mach. & Mech., Sevilla, 425-430, 1987

[16] R. Johanni: **Automatisches Aufstellen der Bewegungsgleichungen von baumstrukturierten Mehrkörpersystemen mit elastischen Bauteilen**, Lst. B f. Mech TUM, Diplomarbeit, 1984

[17] R. Johanni: **On the Automatic Generation of the Equations of Motion for Robots with Elastically Deformable Arms.** Proc. IFAC/IFIP/IMACS Int. Symp. on Theory of Robots, Wien, 1986

[18] R. Johanni: **Optimale Bahnplanung bei Industrierobotern.** Fortschr.-Ber. VDI-Z., Reihe 18, Nr. 51 (1988)

[19] T.R. Kane, R.R. Ryan, A.K. Banerjee: **Dynamics of a Cantilever Beam Attached to a Moving Base.** J. Guidance, Vol. 10, No.2, 139-151, 1987

[20] U. Kleemann: **Dynamics and Control of a Robotic System with Elastic Arms**, Proc. Int. Symp. Robotics and Manipulators, Albuquerque 1986

[21] U. Kleemann: **Regelung elastischer Roboter**, Fortschrittberichte VDI, Reihe 8, Nr. 191, VDI-Verlag, Düsseldorf 1989.

[22] M. Mason: **Compliance and Force Control for Computer Controlled Manipulators.** IEEE Trans. Syst.,Man an Cybernetics, Vol. SMC-11, Nr. 6, 418 - 432, 1981

[23] J.K. Mills, A.A. Goldenberg: **Force and Position Control of Manipulators During Constrained Motion Tasks.** IEEE Trans. Robotics and Automation, Vol. 5, Nr. 1, 1989

[24] S. Nicosia, P. Tomei, A. Tornambe: **Dynamic Modelling of Flexible Manipulators.** Proc. IEEE Conf. on Robotics and Automation, San Francisco/USA, 365 - 372, 1986

[25] F. Pfeiffer, E. Reithmeier: **Roboterdynamik.** Teubner (1987)

[26] F. Pfeiffer, R. Johanni: **A Concept for Manipulator Trajectory Planning.** IEEE J. Robotics and Automation, Vol. RA-3, Nr. 2, 1987

[27] F. Pfeiffer, B. Gebler: **A Multistage Approach to the Dynamics and Control of Elastic Robots.** Proc. IEEE Int. Conf. on Robotics and Automation, Philadelphia/USA, 2 - 8, 1988

[28] F. Pfeiffer, K. Richter: **Optimal Path Planning Including Forces at the Gripper.** To appear in J. of Intelligent and Robotic Systems, Reidel, Holland.

[29] F. Pfeiffer, K. Richter, H. Wapenhans: **Elastic Robot Trajectory Planning with Force Control**, to appear in Proc. of IFIP International Symp. on Theory of Robots, Rom, Italien (1990).

[30] M.H. Raibert, J.J. Craig: **Hybrid Position/Force Control of Manipulators.** Trans. of the ASME, Dyn. Syst., Measurement and Control, Vol. 102, 126 - 133, 1981

[31] K.G. Shi, N.D. McKay: **Minimum-Time Control of Robotic Manipulators with Geometric Path Constraints.** IEEE Trans. Automatic Control 30, 531 - 541, 1985

[32] P. Tomei, S. Nicosia, A. Ficola: **Approach to the Adaptive Control Robots.** Proc. IEEE Conf. on Robotics and Automation, San Francisco/USA, 552 - 558, 1986

[33] A. Truckenbrodt: **Bewegungsverhalten und Regelung hybrider Mehrkörpersysteme mit Anwendung auf Industrieroboter.** Fortschr.-Ber. VDI-Z., Reihe 8, Nr. 33, 1980

[34] A.F. Vereshagin: **Computer Simulation of the Dynamics of Complicated Mechanisms of Robot Manipulators**, Eng. Cybernetics, No. 6, 65-70, 1974

Asymptotic Analysis for Dynamics of Rigid Body Containing Elastic Elements and Viscous Fluid

F. L. CHERNOUSKO

Institute for Problems in Mechanics
the USSR Academy of Sciences,
Moscow, USSR

Summary

We consider motion of a rigid body with internal degrees of freedom: elastic parts and/or tanks containing viscous fluid. If the non-dimensional stiffness and viscosity are sufficiently high then the system is described by singularly perturbed differential equations. By means of asymptotic approach these equations are simplified. Nonlinear evolution of angular motions of the body is analyzed.

Equations of motion for a body with elastic parts

We consider angular motions (with respect to some point 0) of a system which consists of a rigid body R and of an elastic system S attached to it. The system S has n degrees of freedom. Its position with respect to R is defined by generalized co-ordinates $q = (q_1, \ldots, q_n)$. In a neutral position $q = 0$ all elastic elements (springs) of S are not strained. Let J be the tensor of inertia of the system R + S for $q = 0$ with respect to 0 . Denote the angular velocity of the body R and the external torque acting on this body through $\bar{\omega}$ and $\bar{M}$, respectively. The Euler's dynamic and kinematic equations for the body R are [1]

N. V. Banichuk · D. M. Klimov · W. Schiehlen (Eds.)
Dynamical Problems of Rigid-Elastic Systems and Structures
IUTAM Symposium Moscow, USSR, May 23-27, 1990

$$J\dot{\bar{\omega}} + \bar{\omega} \times (J\bar{\omega}) = \bar{M} + \bar{M}', \qquad \bar{M}' = -\dot{\bar{k}} - \bar{\omega} \times \bar{k}$$

$$\bar{k} = \sum_{i=1}^{n} \sum_{j=1}^{3} (\bar{k}_{ij} q_i \omega_j + \bar{k}_i \dot{q}_i), \qquad \dot{\bar{\rho}} = \bar{f}(\bar{\rho}, \bar{\omega}) \tag{1}$$

Here $\bar{\rho}$ is the vector of the Euler angles, the torque $\bar{M}'$ is due to the flexibility of S and vanishes if $q \equiv 0$. Nonlinear (with respect to q, $\dot{q}$) terms in the expression for $\bar{k}$ are omitted since q, $\dot{q}$ are assumed to be small. Vectors $\bar{k}_{ij}$ and $\bar{k}_i$ depend only on geometrical and inertial parameters of the system S and are constant in the co-ordinate system $Ox_1x_2x_3$ attached to R .

Let the system S in its motion with respect to R have the kinetic energy $(A\dot{q}, \dot{q})/2$, the potential elastic energy $(Cq, q)/2$ and the Rayleigh dissipative function $(B\dot{q}, \dot{q})/2$ where A, B and C are constant symmetric positive-definite $n \times n$-matrices. The motion of the system S is described by equations

$$A\ddot{q} + B\dot{q} + Cq = Q \tag{2}$$

where n-vector Q includes external and inertial forces acting on S . The set of equations (1), (2) describing the motion of the system R + S were obtained in [1] where expressions for $\bar{k}_{ij}$, $\bar{k}_i$ are given.

The following assumptions are made

$$B = \varepsilon^{-1} B_0, \qquad C = \varepsilon^{-2} C_0, \qquad 0 < \varepsilon \ll 1 \tag{3}$$

where $B_0 \sim 1$, $C_0 \sim 1$ are constant $n \times n$-matrices, ε is a small parameter. Conditions (3) imply that the maximal period T_0 of elastic vibrations of the system S , time T_d of their damping and the characteristic time T_* of rotation of the body R satisfy the following estimates

$$T_0 \sim T_d \sim \varepsilon T_* , \qquad T_* \sim \omega^{-1} \tag{4}$$

Under the assumptions (3) or (4), the system (1), (2) is reduced to the singularly perturbed system of ordinary differential equations [2].

Singularly perturbed system

General form of such system is

$$\varepsilon \dot{Z} = P(Z, Y, t, \varepsilon), \qquad \dot{Y} = Q(Z, Y, t, \varepsilon) \tag{5}$$

where Z and Y are vectors of fast and slow variables, respectively. Solution $Z(t, \varepsilon)$, $Y(t, \varepsilon)$ of (5) with initial conditions

$$Z(0, \varepsilon) = Z^0, \qquad Y(0, \varepsilon) = Y^0 \tag{6}$$

consists of regular and singular expansions defined by the well known asymptotic procedure [2]. If we are interested only in the slow part of the solution that does not vanish outside the initial boundary layer, then the following result [3] can be useful.

Let all conditions I - IY from [2], pp. 47, 48, 78 be satisfied and functions $H_i(x, t)$ be defined by the formal equality

$$\varepsilon \sum_{i=0}^{\infty} \varepsilon^i \left[\frac{\partial H_i}{\partial t} + \frac{\partial H_i}{\partial x} Q\left(\sum_{i=0}^{\infty} \varepsilon^i H_i, x, t, \varepsilon\right)\right] = P\left(\sum_{i=0}^{\infty} \varepsilon^i H_i, x, t, \varepsilon\right) \tag{7}$$

Functions $\eta_k(t, \varepsilon)$, $\zeta_k(t, \varepsilon)$ are defined by

$$\dot{\eta}_k = Q\left(\sum_{i=0}^{k} \varepsilon^i H_i(\eta_k, t), \eta_k, t, \varepsilon\right)$$

$$\eta_k(0, \varepsilon) = Y^0 + \varepsilon\eta_1^0 + \ldots + \varepsilon^k\eta_k^0 \tag{8}$$

$$\zeta_k(t, \varepsilon) = \sum_{i=0}^{k} \varepsilon^i H_i(\eta_k(t, \varepsilon), t)$$

There exist such η_1^0, η_2^0, ... in (8) that for arbitrary $k = 0, 1, \ldots$ and for sufficiently small ε the following estimates hold ($t_0 > 0$, $C_k > 0$ are constants)

$$|Y(t, \varepsilon) - \eta_k(t, \varepsilon)| \le C_k\varepsilon^{k+1}, \qquad t \in [t_0, T]$$

$$|Z(t, \varepsilon) - \zeta_k(t, \varepsilon)| \le C_k\varepsilon^{k+1} \tag{9}$$

Estimates (9) show that out of the initial boundary layer the singular initial problem (5), (6) can be replaced by the regular initial problem (8) of the lower order. Equations (8) are deduced using only algebraic and differential operations, see (7). However, initial conditions (8) are unknown. They can be either obtained by means of the procedure [2] or, in practical problems, simply measured outside the boundary layer at $t = t_0 = O(1) > 0$. This approach applied to the system (1), (2) under the condition (3) leads to the simplified system for the slow variables $\bar{\omega}$, $\bar{\rho}$ [1, 4].

<u>Euler-Poinsot motion perturbed by internal elastic and dissipative forces</u>

Let no external forces and torques act on the system R + S described above. Then equations for the slow variables $\bar{\omega}$, $\bar{\rho}$ take form [1]

$$J \cdot \dot{\bar{\omega}} + \bar{\omega} \times (J \cdot \bar{\omega}) = \varepsilon^2 \bar{M}_4(\bar{\omega}) + \varepsilon^3 \bar{M}_5(\bar{\omega}), \quad \dot{\bar{\rho}} = \bar{f}(\bar{\rho}, \bar{\omega}) \qquad (10)$$

Equations (10) describe evolution of Euler-Poinsot motion under the influence of internal elasticity and dissipation. Here $\bar{M}_4$ and $\bar{M}_5$ are homogeneous polynomials of the 4th and 5th degrees with respect to components of $\bar{\omega}$

$$\bar{M}_4(\bar{\omega}) = \sum_{i,j,k,p=1}^{3} \bar{D}_{ijkp} \omega_i \omega_j \omega_k \omega_p \qquad (11)$$

$$\bar{M}_5(\bar{\omega}) = \sum_{i,j,k,p,r=1}^{3} \bar{E}_{ijkpr} \omega_i \omega_j \omega_k \omega_p \omega_r$$

Constant coefficients in (11) are expressed through geometrical and mechanical parameters of the system; $\bar{M}_4$ and $\bar{M}_5$ correspond to inertial elasticity and dissipation respectively. Motion of the body subjected to torques (11) tends to the unique stable motion which is a rotation about the axis Oz_1 of the maximal moment of inertia. Characteristic time of this nonlinear transient process is

$$T' \sim \omega^{-4} T_0^4 T_d \qquad (12)$$

Example

Consider a dynamically symmetric rigid body R with a point mass m attached to the point O_1 on its axis by means of a spring with stiffness c and damping coefficient b. Let the body R^* be composed of R and the point mass m placed at O_1. Denote equatorial and axial moments of inertia of the body R^* with respect to its centre of mass O^* by I and I_*, respectively. Let $\bar{G}$ be the angular momentum of the body R^* with respect to O^*.

Consider free motion of the system about its centre of mass (all external forces and torques are zero) under the conditions (3), (4). Then equations (10), (11) have the form ($L = OO^*$).

$$I\dot{\omega}_1 + (I_* - I)\omega_2\omega_3 = \alpha\omega_2\omega_3 + \beta\omega_3^4\omega_1$$

$$I\dot{\omega}_2 + (I - I_*)\omega_1\omega_3 = -\alpha\omega_2\omega_3 + \beta\omega_3^4\omega_2 \tag{13}$$

$$I_*\dot{\omega}_3 = -I\beta\omega_3^3(\omega_1^2 + \omega_3^2)/I_*$$

$$\alpha = m^2L^2c^{-1}I_*I^{-3}G^2, \qquad \beta = m^2L^2c^{-2}bI_*^3(I - I_*)I^{-4}$$

$$G^2 = I^2\omega_1^2 + I^2\omega_2^2 + I_*^2\omega_3^2$$

If follows from (13) that G^2 is constant (in our approximation). Equations (13) were integrated; the result can be expressed as follows

$$\tan^2\theta \exp(\tan^2\theta) = \tan^2\theta_0\exp(\tan^2\theta_0)\exp[(t - t_0)/T'] \tag{14}$$

$$\tan\theta = I(\omega_1^2 + \omega_2^2)^{1/2}(I_*\omega_3)^{-1}, \qquad t_0 = \text{const}$$

$$T' = m^{-2}L^{-2}c^2b^{-1}II_*(I - I_*)^{-1}G^{-4} = \text{const}$$

Here $\theta \in [0, \pi/2)$ is an angle between $\bar{G}$ and the axis of symmetry of the body. The left-hand part of (14) is a monotone function of θ while the right-hand part of (14) is a monotone function of t. Therefore, (14) defines a monotone function $\theta(t)$. If $I > I_*$ then $T' > 0$, θ increases and tends to $\pi/2$ when $t \to \infty$. If $I < I_*$ then $T' < 0$, and θ decreases, $\theta \to 0$ when $t \to \infty$. In both cases, the final motion is a rotation about the axis of the maximal moment of inertia of the body. Some other examples of motions of rigid body with elastic and dissipative elements (both linear and nonlinear) see in [4, 5].

Dynamics of rigid bodies containing viscous or viscoelastic media

Results of the previous sections are extended in [6] to rigid bodies with non-rigid parts made up of continuous viscoelastic medium (Kelvin-Voight material). This medium obeys equations

$$\sigma = G[\sigma'(\bar{u}) + \tau\sigma''(\dot{\bar{u}})], \qquad \sigma' = 2e + 2\mu(1 - 2\mu)^{-1}\text{div}\,\bar{u}E$$

$$\sigma'' = 2\dot{e} - (2/3)\,\text{div}\,\dot{\bar{u}}E, \qquad e_{ij} = (1/2)\left(\frac{\partial u_i}{\partial x_j} + \frac{\partial u_j}{\partial x_i}\right) \tag{15}$$

$$i, j = 1, 2, 3$$

where σ is the stress tensor made up of elastic and viscoelastic stresses; G is the shear modulus; μ is the Poisson ratio; τ is a constant with the dimension of time (shear relaxation time); e is the strain tensor; $\bar{u}$ is the displacement vector; x_i are Cartesian co-ordinates associated with the rigid body; and E is the unity tensor. Partial differential equations for the viscoelastic medium replace here ordinary differential equations (2). Equations of motion include these partial differential equations and also ordinary differential equations for the rigid body. Assumptions (3) are replaced here by

$$G = \varepsilon^{-2}G_0, \qquad \tau = \varepsilon T_*, \qquad T_* \sim \omega^{-1}, \qquad 0 < \varepsilon \ll 1 \qquad (16)$$

Relations (16) lead to the estimates (4) equivalent to:

$$v_0 \sim L(T_d)^{-1} \gg v^*, \qquad v^* = \omega L$$

where v_0 is the characteristic speed of elastic waves; v^* is the linear rotational velocity of the body; and L is its linear dimension. Under conditions (16) or (4), free elastic vibrations are damped rapidly and can be disregarded for $t \gg T_d \sim \varepsilon$. Asymptotic analysis of equations of motion (both partial and ordinary differential equations) leads to simplified ordinary differential equations for slow variables similar to that obtained for a system with a finite number of degrees of freedom. For instance, if the body is free (all external forces and torques are zero) then we have the same equations (10), (11). Time of transient processes is given by (12), or $T' \sim \omega^{-4}T_0^{-2}\tau^{-1}$.

Compare these results with the case of a rigid body with a cavity containing viscous fluid with small Reynolds number $Re = \omega L^2\nu^{-1} \ll 1$. Here L is the linear dimension of the cavity, and ν is the kinematic viscosity of the fluid. It was shown in [7] that equations of free motion of the body about its centre of mass can be written as

$$J\cdot\dot{\bar{\omega}} + \bar{\omega} \times (J\cdot\bar{\omega}) = Re\ \bar{M}_3(\bar{\omega}), \qquad Re \ll 1$$

$$\bar{M}_3(\bar{\omega}) = \sum_{i,j,k=1}^{3} \bar{F}_{ijk}\omega_i\omega_j\omega_k \tag{17}$$

Here $\bar{M}_3(\bar{\omega})$ is a homogeneous polynomial of the 3d degree. Its coefficients as well as similar coefficients in (11) are constant and depend only on the geometry of the cavity. To find the coefficients in (11), (17), one needs to solve certain stationary boundary value problems for viscoelastic or viscous medium.

The case of large Reynolds number ($Re \gg 1$) is quite different. This case was studied in [8] for small laminar motions of the viscous fluid in a cavity; here asymptotic approach was applied as well. A number of results on dynamics of a rigid body with a cavity containing viscous fluid can be found in the book [9].

Conclusion

Results summed up in this paper make it possible to analyze angular motions of artificial satellites and celestial bodies under the influence of small internal perturbation torques. Asymptotic approach permits to obtain some qualitative results and to describe evolution of motions using simplified averaged equations and avoiding numerical integration of fast motions with small step needed in a direct approach. Results obtained in [1, 3-10] made it possible to estimate influence of the internal flexibility and dissipation on satellite angular motions, to evaluate dynamical effects caused by the presence of flexible elements and fluid in tanks, to choose parameters of dampers atc.

References

1. Chernousko, F.L.: On motion of a solid body with elastic and dissipative elements. Appl. Math. and Mech. (PMM) 42 (1978) 32-41.

2. Vasilyeva, A.B.; Butuzov, V.F.: Asymptotic expansions for solutions of singularly perturbed systems. Moscow: Nauka 1973 (In Russian).

3. Chernousko, F.L.; Shamayev, A.S.: Evolutionary equations for slow variables in the theory of singularly perturbed systems. Doklady of USSR Acad. of Sci.277 (1984) 315-318.

4. Chernousko, F.L.: On the motion of rigid body with moving internal masses. Mech. of Solids (Izv. AN SSR, MTT) 8, N 4 (1973) 33-44.

5. Akulenko, L.D.; Leshchenko D.D.: Some problems of motion of rigid body with a movable mass. Mech. of Solids (Izv. AN SSR, MTT) 13, N 3 (1978) 29-34.

6. Chernousko, F.L.: Motion of a viscoelastic solid relative to the centre of mass. Mech. of Solids (Izv. AN SSR, MTT) 15, N 1 (1980) 17-21.

7. Chernousko, F.L.: Motion of a rigid body with cavities filled with a viscous fluid at low Reynolds numbers. J. of Comput. Math. and Math. Physics (Zh. Vychisl. Matem. i Matem. Fiz.) 5 (1965) 1049-1070.

8. Chernousko, F.L.: Motion of a body with a cavity filled with a viscous fluid at large Reynolds numbers. Appl. Math. and Mech. (PMM) 30 (1966) 568-589.

9. Chernousko, F.L.: Motion of a rigid body with cavities containing viscous fluid. Moscow: Computing Centre of USSR Acad. of Sci. 1968 (in Russian).

10. Chernousko, F.L.: Asymptotic methods in dynamics of systems with elastic and dissipative elements. Proc. 4th Intern. Conf. on Boundary and Interior Layers (BAIL IY). Dublin: 1986, 15-26.

Stability of Vibratory Modes in Moving Media

Stephen H. Crandall

Massachusetts Institute of Technology
Cambridge, Massachusetts 02139, U.S.A.

SUMMARY

In many mechanical systems a uniform linear or rotational motion is imposed on part of the system; e.g., rail vehicles, bandsaws, turbines, pumps, motors, generators. In such systems the kinetic energy of the steady motion provides a large reservoir of energy which under certain circumstances can be channelled into unstable vibratory modes. This paper describes a variety of destabilizing mechanisms which permit the conversion of the steady-motion energy into vibratory energy in uniaxial wave-propagation subsystems in relative longitudinal motion and in various rotordynamic systems.

INTRODUCTION

A mechanical system in which one part is in uniform translational or rotational motion with respect to the rest has interesting dynamic properties. In general the system as a whole is nonconservative. The small motion linear vibration behavior of the system generally depends on the speed of the moving part. In particular the natural frequencies and loss factors of the vibration modes can vary with the speed. Most dramatically a stable dynamic system can be transformed into an unstable system by changing the speed. The purpose of the present paper is to provide simple physical explanations of the mechanisms responsible for destabilizing such systems. A common analytical approach is to assign e^{st} time behavior to the dynamical coordinates and use the linearized equations of motions to study the root loci as the speed is varied. In a stable lightly damped system the roots tend to be distributed close to the imaginary axis of the complex s-plane. For a stable system all roots lie in the left half-plane. As the speed of the moving part is varied the roots migrate. Instability results when one or more roots cross into the right half-plane. There appear to be two major patterns of root movement leading to instability. In the first, a particular root, accompanied by its complex conjugate, moves to the right and enters the right half-plane. The speed change here is primarily altering the damping of the mode without significantly altering the frequency. In the second pattern, a *pair* of roots, accompanied by their complex conjugates, move toward one another along paths which are close to, and essentially parallel to, the imaginary s-axis. Here the speed change is primarily altering the stiffness and/or inertia of the modes which is reflected in natural frequency changes. As the pair

N. V. Banichuk · D. M. Klimov · W. Schiehlen (Eds.)
Dynamical Problems of Rigid-Elastic
Systems and Structures
IUTAM Symposium Moscow, USSR, May 23-27, 1990

of roots approach one another the natural frequencies of two distinct modes tend to the same value. Then depending on the nature of the coupling between the modes [1] the roots either repel one another if the coupling is conservative, or attract one another if the coupling is nonconservative. In the former case the roots remain in the left half-plane and the stability of the system is unchanged. In the latter case the roots suddenly swerve at or slightly before the point of collision and with continued speed change they separate along paths at right angles to their direction of approach. One root of the pair moves further left into the left half-plane while the other moves into the right half-plane triggering instability. Physical mechanisms which exhibit these two patterns are described for simplified systems with long flexible members in longitudinal motion which represent idealizations of actual systems containing moving belts or bandsaw blades. Reference is also made to analogous behavior in rotordynamic systems.

DESTABILIZATION DUE TO DAMPING APPLIED TO A MOVING ELEMENT

Consider as an idealization of a moving belt an infinitely long uniform flexible member traveling longitudinally with velocity v along the x-axis. In general the elastic restoring forces tending to keep the belt straight result from some particular combination of bending stiffness EI, belt tension T, and elastic foundation modulus α. It is assumed that the belt axis remains in the plane $y = 0$ but may make small transverse deflections $w(x,t)$ in the z-direction. The mass per unit length of the belt is ρA. For simplicity rotatory inertia is neglected. In the simple case where only belt tension T is active the belt supports forward and backward waves of small transverse displacement as free undamped vibrations. The phase velocity of these waves, as measured by an observer on the belt, is c where

$$c^2 = \frac{T}{\rho A} \tag{1}$$

An observer in a fixed reference system would observe the phase velocity of the forward wave as $c + v$ and the phase velocity of the backward wave as $-(c - v)$. Now consider that a uniformly distributed transverse viscous damping is added to the system as indicated in Fig. 1. Let the damping force per unit length acting on the belt be $-b\partial w/\partial t$. The primary effect of this damping, at least for small v, is to introduce temporal decay into the free vibration forward and backward waves. For sinusoidal waves of fixed wavenumber k the introduction of first-order small damping b produces first-order temporal decay rate but only second order change in natural frequency. The rate of decay, however, depends on the belt speed v. For the forward wave the decay rate increases with increasing v while for the backward wave the decay rate decreases with increasing v. As v is increased further the decay rate for the backward wave drops to zero and changes sign to become an amplification rate. It will be shown that the changeover from decay to amplification occurs when the phase velocity of the backward wave as observed from the fixed reference system is zero; i.e., when the transport velocity v is just equal and opposite to the phase velocity c within the moving medium. When $v > c$, and the backward wave is being dragged forward faster than it propagates through the medium, the external damping amplifies the initial disturbance and the system is unstable.

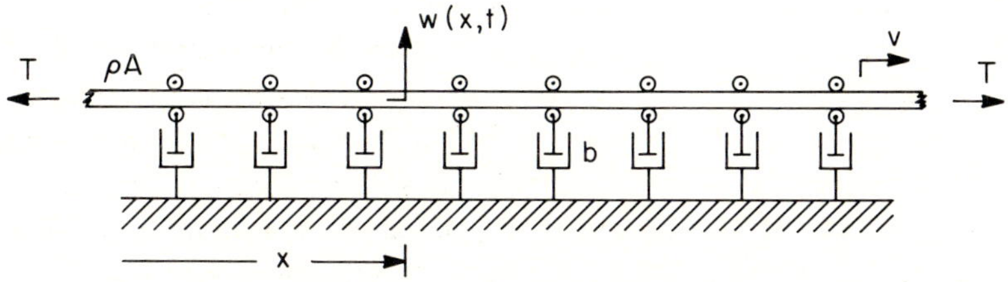

Fig. 1. Moving belt subjected to fixed viscous damping.

The crux of the argument lies in the kinematics of the moving belt. The transverse displacement $w(x,t)$ of the belt is measured with respect to the fixed reference system. At an arbitrary location x the velocity of the local dashpot terminal is $\partial w/\partial t$ so that the transverse dashpot force per unit length acting on the belt is $-b\partial w/\partial t$. This force is transmitted through ideal rollers to the belt which has a longitudinal transport velocity v as well as a small transverse velocity. At the location x the transverse velocity component of the belt is given by the convective derivative

$$\frac{Dw}{Dt} = \frac{\partial w}{\partial t} + v\frac{\partial w}{\partial x} \tag{2}$$

The rate of work done on the belt by the dashpots, per unit length of belt, is the power flow

$$P = -b\frac{\partial w}{\partial t}\frac{Dw}{Dt} \tag{3}$$

When P is negative, energy is being extracted from the transverse motion of the belt. Consider next a forward wave component with arbitrary wave number k which at a certain point during the decay process has the form

$$w = a\sin k[x - (c+v)t] \tag{4}$$

For this component the average rate of energy transfer to this component per unit length of belt is

$$< P >= -\frac{1}{2}b\,a^2k^2c(c+v) \tag{5}$$

which is intrinsically negative for $v > 0$. The dashpots remove energy from forward wave components of arbitrary wavenumber. For a backward wave component

$$w = a\sin k[x + (c-v)t] \tag{6}$$

however, the corresponding calculation leads to

$$< P >= -\frac{1}{2}b\,a^2k^2c(c-v) \tag{7}$$

which shows that the energy removed from backward wave components decreases with increasing v until $v = c$. For $v > c$ the power flow reverses direction and the dashpots actually deliver energy to the backward wave components increasing the amplitude of the disturbance.

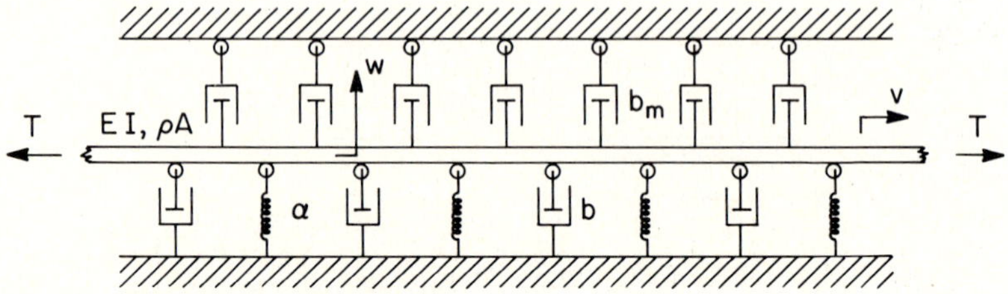

Fig. 2. Moving belt subjected to both fixed and moving viscous damping.

A second explanation follows from a formal analysis based on the equation of motion for the moving belt. Here the system is generalized in Fig. 2 to include bending stiffness EI and an elastic foundation with modulus α in addition to the belt tension T as elastic restoring elements. Also in addition to the stationary viscous damping with force per unit length $-b\partial w/\partial t$, a moving viscous damping with force per unit length $-b_m Dw/Dt$ is included. The equation of motion for unforced small transverse motions of the moving belt with respect to the fixed reference frame is

$$EI\frac{\partial^4 w}{\partial x^2} - T\frac{\partial^2 w}{\partial x^2} + \alpha w + b\frac{\partial w}{\partial t} + b_m\frac{Dw}{Dt} + \rho A\frac{D^2 w}{Dt^2} = 0 \tag{8}$$

where the convective time derivative D/Dt is defined by (2). Wavelike solutions with wavenumber k are sought in the form

$$w = a\, e^{i(kx-\omega t)} \tag{9}$$

The stability of the solution depends on the imaginary part of ω. If $I_m\{\omega\} > 0$ the solution (9) grows exponentially in time indicating instability. Substitution of (9) into (8) yields an algebraic equation for the determination of ω. It is convenient here to introduce the phase velocity c_0 of the undamped ($b = b_m = 0$) stationary ($v = 0$) belt, given by

$$c_0^2 = \frac{EIk^4 + Tk^2 + \alpha}{\rho A k^2} \tag{10}$$

and to introduce alternative damping parameters with the dimension of frequency given by

$$2\beta = \frac{b}{\rho A} \qquad\qquad 2\beta_m = \frac{b_m}{\rho A} \tag{11}$$

The equation for ω then becomes

$$\omega^2 - 2\omega(kv - i\beta - i\beta_m) - k^2(c_0^2 - v^2) - i\beta_m 2kv = 0 \tag{12}$$

Its roots are

$$\omega = kv - i(\beta+\beta_m) \pm kc_0\left[1-\left(\frac{\beta+\beta_m}{kc_0}\right)^2 - i\frac{2\beta}{kc_0}\cdot\frac{v}{c_0}\right]^{1/2}$$
$$= k(v \pm c_0\gamma) - i\left[\beta_m + \beta\left(1 \pm \frac{v}{\gamma c_0}\right)\right] \tag{13}$$

where γ is the positive root of

$$\gamma^2 = \frac{1}{2}\left\{1-\left(\frac{\beta+\beta_m}{kc_0}\right)^2 + \left[\left(1-\left\{\frac{\beta+\beta_m}{kc_0}\right\}^2\right)^2 + \left(\frac{2\beta}{kc_0}\cdot\frac{v}{c_0}\right)^2\right]^{1/2}\right\} \tag{14}$$

For light damping γ may be taken as unity. It can be seen from (13) that the upper signs refer to a forward wave with phase velocity $c_0\gamma + v$ and with a decay rate which increases with the transport speed v. The lower signs refer to a backward wave with phase velocity $c_0\gamma - v$ with a decay rate which decreases with the transport speed v. The backward wave becomes unstable when

$$v = c_0\gamma\left(1+\frac{\beta_m}{\beta}\right) \tag{15}$$

This result indicates that the moving damping is always stabilizing and delays the onset of instability due to the stationary damping.

These results for transverse wave motions on moving belts are similar to those for whirling modes in rotordynamic systems except for the reversal of the roles of stationary and moving damping. Kimball [2,3] showed that damping in the rotor is destabilizing when the rotational speed is greater than the natural frequency of whirling. Smith [3] considered viscous damping both in the rotor and externally and showed in a result analagous to (15) that external damping delayed the onset of instability due to internal damping.

DESTABILIZATION DUE TO MODE COUPLING

To provide a simple example of mode coupling consider a pair of elastically coupled flexible belts as shown in Fig. 3. Let the transverse displacements of the belts as seen

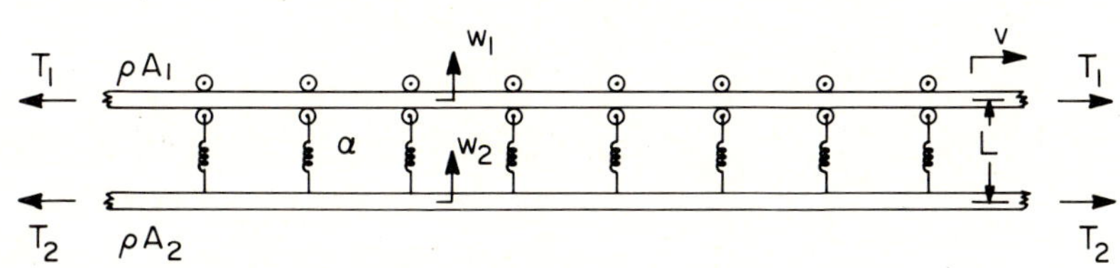

Fig. 3. Moving belt elastically coupled to stationary belt.

by a stationary observer be $w_1(x,t)$ and $w_2(x,t)$. The elastic coupling force per unit length tending to restore the separation distance L is taken to be $\alpha(w_1 - w_2)$. If, for simplicity, damping is neglected and the only elastic restoring agencies considered in the belts are those due to the belt tensions T_1 and T_2, the equations of motion for the coupled pair of belts are

$$\begin{aligned} -T_1\frac{\partial^2 w_1}{\partial x^2} + \rho A_1\frac{D^2 w_1}{Dt^2} + \alpha(w_1 - w_2) = 0 \\ -T_2\frac{\partial^2 w_2}{\partial x^2} + \rho A_2\frac{\partial^2 w_2}{\partial t^2} - \alpha(w_1 - w_2 = 0 \end{aligned} \tag{16}$$

where the convective derivative (2) is used to account for the transport velocity v of the first belt. Coupled wave solutions with wavenumber k are sought in the form

$$\begin{aligned} w_1 = a_1 e^{i(kx-\omega t)} \\ w_2 = a_2 e^{i(kx-\omega t)} \end{aligned} \tag{17}$$

It is convenient here to introduce the individual phase velocities c_1 and c_2 given by

$$c_1^2 = \frac{T_1 k^2 + \alpha}{\rho A_1 k^2} \qquad c_2^2 = \frac{T_2 k^2 + \alpha}{\rho A_2 k^2} \tag{18}$$

and alternative coupling parameters, ϵ_1 and ϵ_2, having the dimension of frequency, given by

$$\epsilon_1^2 = \frac{\alpha}{\rho A_1} \qquad \epsilon_2^2 = \frac{\alpha}{\rho A_2} \tag{19}$$

Then substitution of (17) into (16) leads to the eigenvalue problem

$$\begin{bmatrix} k^2 c_1^2 - (\omega - kv)^2 & -\epsilon_1^2 \\ -\epsilon_2^2 & k^2 c_2^2 - \omega^2 \end{bmatrix} \begin{Bmatrix} a_1 \\ a_2 \end{Bmatrix} = 0 \tag{20}$$

for the frequency ω. For each value of the transport speed v there are four roots for ω. For weak coupling; i.e., small off-diagonal terms in (20), the coupled mode shapes $\{a_1\ a_2\}^T$ are generally very nearly of the form $\{1\ 0\}^T$ or $\{0\ 1\}^T$ with ω-roots very close to the separate roots of the diagonal elements of (20). The exceptional cases in which there is strong interaction between the belts occur in the neighborhoods of the conditions for which both belts have identical uncoupled natural frequencies. The uncoupled natural frequencies for belts 1 and 2

$$\omega_1 = kv \pm kc_1 \qquad \text{and} \qquad \omega_2 = \pm kc_2 \tag{21}$$

are displayed in Fig. 4. The coupling introduced by the off-diagonal elements in (20) may be either conservative or nonconservative [1]. In the former case there are always two distinct modes and the coupled natural frequencies repel each other. In the latter case the modes coalesce and then there is a range of speeds for which the coupled natural frequencies have the same real part but with imaginary parts of opposite sign. The nature of the coupling can be determined by solving the eigenvalue problem (20) exactly for the natural frequencies as functions of the transport speed v. In the case of

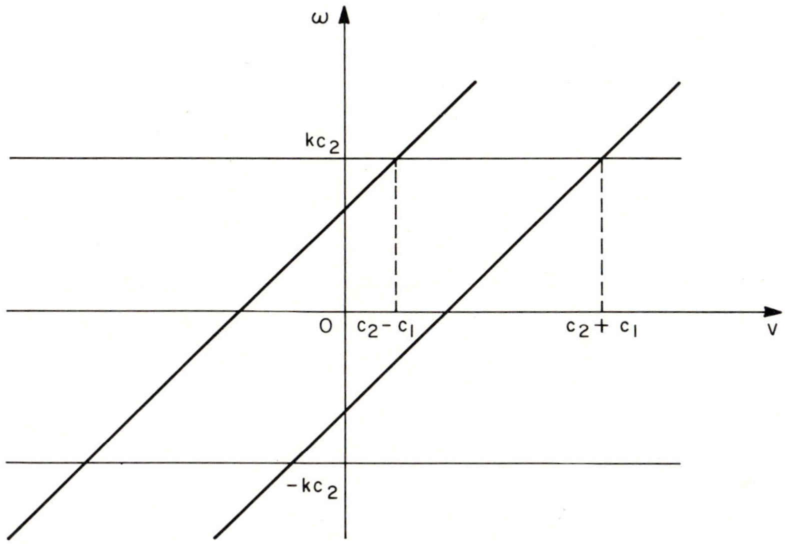

Fig. 4. Uncoupled natural frequencies of pair of belts. Note four real roots for each transport speed v.

weak coupling a simple perturbation calculation is adequate. For example, to investigate the effect of coupling in the neighborhood of $v = c_2 - c_1$ set

$$\omega = kc_2 + \Delta \tag{22}$$

in (20). Neglecting terms proportional to Δ^2 in comparison with terms proportional to Δ one finds

$$\Delta = \pm \frac{\epsilon_1 \epsilon_2}{2k(c_1 c_2)^{1/2}} \tag{23}$$

indicating a pair of distinct real frequencies for the coupled system at the speed for which the uncoupled modes have the same frequency. Similarly, to investigate the effect coupling in the neighborhood of $v = c_2 + c_1$, again take ω in the form (22) in (20). Neglecting Δ^2 terms one finds

$$\Delta = \pm i \frac{\epsilon_1 \epsilon_2}{2k(c_1 c_2)^{1/2}} \tag{24}$$

indicating a pair of complex conjugate natural frequencies. The same elastic coupling mechanism between a stationary belt and a moving belt thus provides both conservative and nonconservative coupling as shown in Fig. 5. In the neighborhood of $v = c_2 - c_1$ when the transport velocity $c_2 - c_1$ boosts the slower phase velocity c_1 in the moving belt to match the phase velocity c_2 in the stationary belt the coupling is conservative and the coupled motions are stable. On the other hand, in the neighborhood of $v = c_2 + c_1$ when the transport velocity overcomes the backward phase velocity $-c_1$ in the moving belt and drags it forward to match the phase velocity c_2 in the stationary belt the coupling is nonconservative and the coupled motions are unstable.

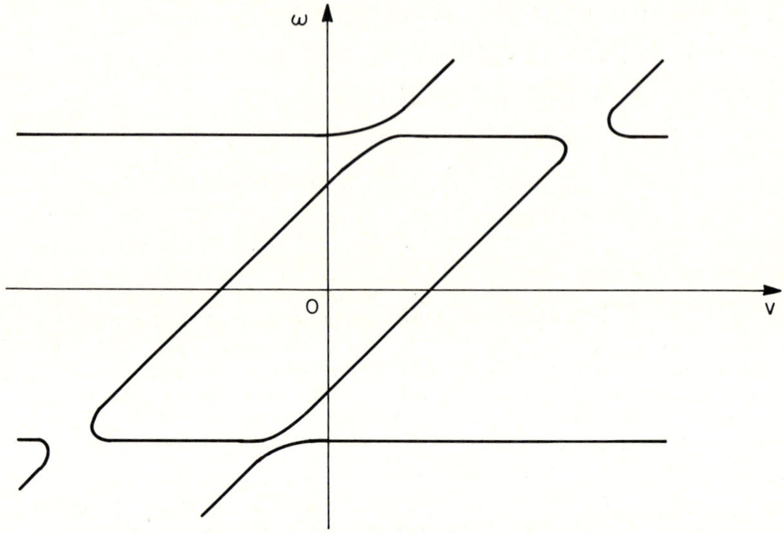

Fig. 5. Coupled natural frequencies. Complex conjugate roots occur in gaps where only two real roots exist.

Similar coupling stability phenomena occur in rotordynamic systems [1]. In axially symmetric systems circular whirling motions of the stator are possible. In rotors with propeller blades or with cavities partially filled with liquid, whirling modes in the rotor are also possible. The stator and rotor whirling modes are inertially coupled. The whirl rate of a rotor whirl mode as observed by a stationary observer depends on the rotation rate of the rotor. As in the case of coupled belts strong interaction between moving and stationary modes occurs only in the neighborhoods where the uncoupled natural frequencies are the same. The coupling is nonconservative and the system unstable when the equality of whirl frequency is obtained by rotating a backward rotor whirl so fast that it becomes a forward whirl at the same frequency as a forward stator whirl.

REFERENCES

1. Crandall, S.H., and Mroszczyk, J.W., Conservative and nonconservative coupling in dynamic systems, pp. 567–572 in Proc. I. Mech. E. Intl. Conf. 1988-7, *Vibrations in Rotating Machinery*, I. Mech. E., London, 1988.
2. Kimball, A.L., Jr., Internal friction theory of shaft whirling, *Phys. Rev.* **2**21, 703 (1923).
3. Kimball, A.L., Jr., Internal friction as a cause of shaft whirling, *Phil. Mag.* **6**49, 724–727 (1925).
4. Smith, D.M., The motion of a rotor carried by a flexible shaft in flexible bearings, *Proc. Roy. Soc. London* A142, 92–118 (1933).

A Numerical Integration Algorithm Based on the Introduction of a Constraint Violation Factor

DEHOMBREUX P.; CONTI C.; VERLINDEN O.

Service de Mécanique Rationnelle, Faculté Polytechnique de Mons
Boulevard Dolez, 31 7000 Mons Belgium

Introduction

The direct dynamic solution approach developed here for closed-chain mechanisms is based on Lagrangian formalism, relative coordinates and constraint equations derived from loop-closure conditions. In this way, the dynamical problem is formulated as a system of Lagrange differential equations in the relative coordinates, accompanied by second-order constraints. These equations are linear in the second-order terms and are solved by a numerical integration procedure. Integration algorithms determine numerical stability and global features of direct dynamic analysis in a large extent.

This paper propounds the introduction of a "Constraint Violation Factor" improving the efficiency of the integration stage. A technique for eliminating some of the drawbacks of the well-known Baumgarte's stabilization method is presented.

1. Dynamics of Kinematically Constrained Systems.

A mechanical system may contain one or more independent closed kinematic loops. For example, the system shown in Fig. 1(a) contains two closed loops. If the closed loops are cut at one of their joints, as in Fig. 1(b), the mechanism becomes equivalent to an open-chain system.

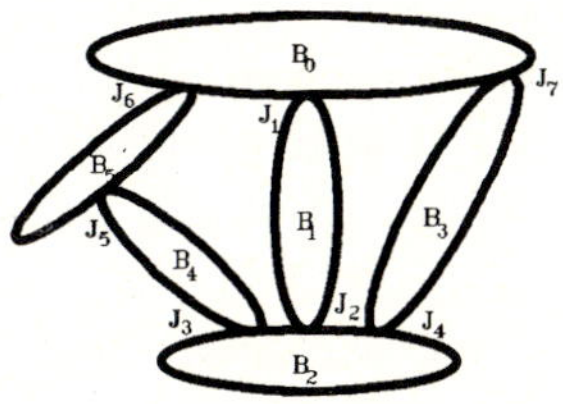

A system containing two closed loops

(Fig. 1.a)

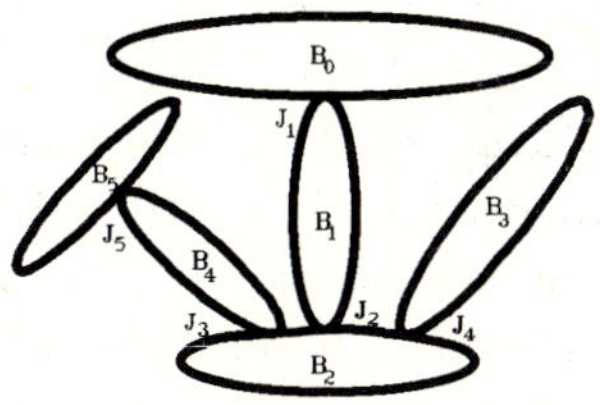

The equivalent open-chain system

(Fig. 1.b)

The differential motion equations of the so-called "equivalent open-chain system" are formulated by following Lagrange's equation (of second type) in the relative coordinates.

$$\frac{d}{dt}\left(\frac{\partial L}{\partial \dot{q}_j}\right)-\frac{\partial L}{\partial q_j}=Q_j, \qquad L=K-P \qquad \text{(Eq.1.a)} \quad \text{(Eq.1.b)}$$

with K = total system kinetic energy,
P = energy supplied to the conservative fields by the system,
L = Lagrangian,
q_j = relative coordinates (Lagrangian coordinates).
Q_j = generalized force associated with coordinate q_j.

N. V. Banichuk · D. M. Klimov · W. Schiehlen (Eds.)
Dynamical Problems of Rigid-Elastic Systems and Structures
IUTAM Symposium Moscow, USSR, May 23-27, 1990

The application of the Lagrange's equations will require an expression for kinetic and potential energies as a function of the relative coordinates. Finally, the equations of motion can be written in matrix notation as (cf. ref. [2])

$$[M(\{q_j\})]\{\ddot{q}_j\}+\{V(\{q_j\},\{\dot{q}_j\})\}+\{G(\{q_j\})\}=\{Q_j\} \qquad \text{(Eq.1.c)}$$

with $[M(\{q_j\})]$ = matrix of inertia properties.
$\{V(\{q_j\},\{\dot{q}_j\})\}$ = vector of centrifugal and Coriolis terms
$\{G(\{q_j\})\}$ = vector of gravity terms
$\{Q_j\}$ = vector of forces directly applied to the joints.

With the aim of obtaining the motion equations of the closed mechanism, as many constraint equations as degrees of freedom produced by the different joint cuts must be introduced. These equations are written implicitly in the form:

$$f_i(\{q_j\})=\{0\} \qquad (\mathrm{i} = 1 \ldots \mathrm{N}) \qquad \text{(Eq.1.d)}$$

These N constraint equations $f_i(\{q_j\})=\{0\}$ express dependencies between the M generalized coordinates q_j, the mechanism being characterized by F = M - N degrees of freedom.

Any mechanism can be considered as a series of bodies (links) connected together by joints. Different coordinates frames can be rigidly attached to each body. Using transformation matrices, the relative position and orientation between two coordinate frames can be described. T_1 gives position and orientation of the first link, T_2 gives the position and orientation of the second link with respect to the first, and so forth. Considering the fact that each kinematic loop forms a closed sequence of coordinates frames, it is possible to write in symbolic notation that:

$$T_1(q_1).T_2(q_2).....T_M(q_M).T_{M+1}=I \qquad \text{(Eq.1.e)}$$

The symbol I designing the identity matrix indicates that the loop is closed. All the successive transformations necessary to close the loop apply a coordinate frame on itself.

These constraint equations can be written in their variational form:

$$\delta f_i=\sum_{j=1}^{M} \frac{\partial f_i}{\partial q_j} \delta q_j=0 \qquad \text{(Eq.1.f)}$$

This last equation defines the Jacobian matrix $[C]$. The kinematic constraints are:

$$[C]_{N\times M} \ \{\dot{q}_j\}=\{0\} \qquad \text{(Eq.1.g)}$$

Derivative operators can be used to achieve partial differentiation of transformation matrices with respect to their own pair variables.

The velocities constraints (Eq.1.g) abovementioned are differentiated one more time:

$$[C] \ \{\ddot{q}_j\}=-[\dot{C}]\{\dot{q}_j\} \qquad \text{(Eq.1.h)}$$

in order to obtain a single system in the accelerations $\{\ddot{q}_j\}$, represented in matrix form as

$$\begin{bmatrix} [M(\{q_j\})] & \left[\frac{\partial f_i}{\partial q_j}\right]^T \\ \left[\frac{\partial f_i}{\partial q_j}\right] & [0] \end{bmatrix} \begin{Bmatrix} \{\ddot{q}_j\} \\ \{\lambda_i\} \end{Bmatrix} = \begin{Bmatrix} \{Q_j\} - \{V(\{q_j\},\{\dot{q}_j\})\} - \{G(\{q_j\})\} \\ -\left[\sum\sum \frac{\partial^2 f_i}{\partial q_k \partial q_l} \dot{q}_k \dot{q}_l\right] \end{Bmatrix} \quad \text{(Eq.1.i)}$$

A set of N Lagrange multipliers has been introduced, in order to take into account forces exteriorized by the joint cuts. They can be interpreted as reactions produced by the constraints as external forces.

The dynamic problem has been formulated as a simultaneous set of Lagrange differential equations in the system relative coordinates, accompanied by second-order constraints. These equations are linear in accelerations $\ddot{q}_j$. A direct integration algorithm (for instance, as the Newmark's scheme) can be used to get velocities $\dot{q}_j$ and positions q_j .

2. Constraint Violation Stabilization Method.

The motion of the system, under the action of applied forces, is required to be consistent with geometric and kinematic relations that are imposed on the system by the loop-closure conditions. The accumulation of numerical error may lead to substantial violation of the position and velocity constraint equations.

In other words, it is possible to satisfy fully the equations:

$$\left\{\frac{\partial^2 f_i(\{q_j\})}{\partial t^2}\right\} = [C]\{\ddot{q}_j\} + [\dot{C}]\{\dot{q}_j\} = \{0\} \quad \text{(Eq.2.a)}$$

without verifying nor the kinematic constraints (in the velocities) nor the geometric constraints (in the positions) :

$$\left\{\frac{\partial f_i(\{q_j\})}{\partial t}\right\} = [C]\{\dot{q}_j\} = \{0\} \quad \text{(Eq.2.b)}$$

$$\{f_i(\{q_j\})\} = \{0\} \quad \text{(Eq.2.c)}$$

Estimations of velocities and positions obtained by the integration scheme are inevitably different from exact ones. The violations of constraints are denoted as:

$$\left\{\frac{\partial f_i(\{q_j\})}{\partial t}\right\} = [C]\{\dot{q}_j\} = \{\epsilon_i\} \quad \text{(Eq.2.d)}$$

$$\{f_i(\{q_j\})\} = \{\theta_i\} \quad \text{(Eq.2.e)}$$

In the next integration step, the constraint violations are:

$$\{f_i(\{q_j\})\} = \{\theta_i\} + h\{\epsilon_i\} \quad \text{(Eq.2.f)}$$

The errors propagate as the simulation proceeds and can be considerably amplified. In practice, the conditions (Eq.2.b) and (Eq.2.c) can be slightly transgressed. Nevertheless, the phenomenon has to be controlled, in order to avoid inordinate errors and divergence.

In control theory, it is well known that systems described by second-order differential equations such as $\ddot{y}=0$ are unstable, since noise (or numerical error, in the case of a integration algorithm) may be amplified. However, systems described by differential equations such as $\ddot{y}+2a\dot{y}+b^2y=0$ in which $a,b>0$ are stable.

Following this idea, Baumgarte has shown (cf. ref. [1]) how the numerical accuracy of the solutions of differential equations of motion can be improved. In order to reduce the constraint violation at the cut edges, constraint stabilization can be introduced. Appending stabilization terms $2a\dot{f}_i$ and $b^2 f_i$ in

$$\left\{\frac{\partial^2 f_i(\{q_j\})}{\partial t^2}\right\}=\{0\} \tag{Eq.2.g}$$

yields the following equation:

$$\left\{\frac{\partial^2 f_i(\{q_j\})}{\partial t^2}\right\}+2a\left\{\frac{\partial f_i(\{q_j\})}{\partial t}\right\}+b^2\{f_i\}=0 \qquad \text{with} \quad a,b>0 \tag{Eq.2.h}$$

In practice, the values a and b range from 0 to 50. When a equals b, critical damping is achieved (Cf. Ref. [1],[2],[3])

The stabilized formulation of motion equations is:

$$\begin{bmatrix}[M] & [C]^T \\ [C] & [0]\end{bmatrix}\begin{Bmatrix}\{\ddot{q}_j\} \\ \{\lambda_i\}\end{Bmatrix}=\begin{Bmatrix}\{Q\} \\ -[\dot{C}]\{\dot{q}_j\}-2a[C]\{\dot{q}_j\}-b^2\{f_i\}\end{Bmatrix} \tag{Eq.2.i}$$

If, on the one hand, the additional computations involved in the evaluation of the stabilization terms can be considered as negligible, on the other hand, to satisfy the convergence test in the predictor-corrector scheme, the number of iterations increases.

If the stabilization terms a and b are overestimated, non-convergence of integration can arise. If the mechanism reach a kinematically singular position, the jacobian matrix is rank deficient and a small perturbation of the right side of equations (2.i) can provide large error in solution: the stabilization terms become prevailing and the integration algorithm may fail to converge.

3. The Constraint Violation Factor.

With the intent to investigate the constraint violation phenomenon, a constraint violation factor, CVF, has been defined.

Ideally, the matrix $[I^*]$ defined as the ordered product of the transformations matrices describing each closed kinematic chain would be equal to the unit matrix I. The CVF_g will measure the differences between the two matrices.

Among the components of a 4x4 homogeneous matrix, it is advisable to distinguish the 3x3 matrix of direction cosines and the 3x1 vector defining respectively relative orientation and position of two coordinates frames. This distinction induces us to define two geometrical constraint violation factors: one, CVF_g^o, with respect to orientation; the other, CVF_g^p, with respect to position.

$$CVF_g^o = \sqrt{\sum_{l=1}^{3}\sum_{c=1}^{3}(I^*_{l,c} - I_{l,c})^2} \qquad \text{(Eq.3.a)}$$

$$CVF_g^p = \sqrt{\sum_{l=1}^{3}(I^*_{l,4})^2} \qquad \text{(Eq.3.b)}$$

Ultimately, the Constraint Violation Factor is built from both. A factor k has been introduced, in order to satisfy the dimensional homogeneity of equation 3.c. In practice, the evolution of CVF_g^p is sufficiently significant: a zero value can be taken for the dimensional factor k.

$$CVF_g = CVF_g^p + k\ CVF_g^o \qquad \text{(Eq.3.c)}$$

4. The Constraint Violation Factor... a Quality Index.

An absence or a deficiency of error control on the constraints violation may lead to erroneous results. The user is able to monitor global constraint error by referring to the CVF curve. If an unacceptable constraint violation is encountered, the user has to define other parameters for the integration process, for, instance, by reducing the step size.

The evolution of CVF_g during the simulation of the multibody system shown in figure 4.a (7 revolute joints, 5 links, 2 loops) has been represented hereunder (Cf. fig. 4.b).

Different curves have been obtained for three distinct sets of integration parameters:
($a = b = 0$, $\Delta t = 0.025$s [A]),
($a = b = 10$, $\Delta t = 0.025$s [B]) and
($a = b = 10$, $\Delta t = 0.005$s [C]).

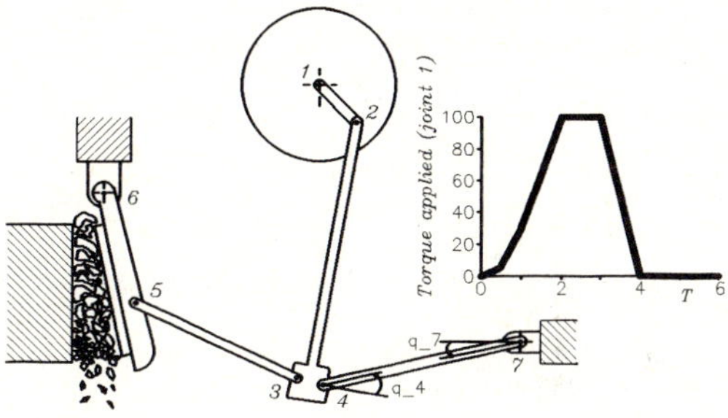

(Fig. 4.a)

The use of a so-called direct integration method without any form of stabilization leads to a increasing violation of the geometric constraints (curve A): if a graphic animation of the mechanism is displayed, the linkage will appear disconnected.

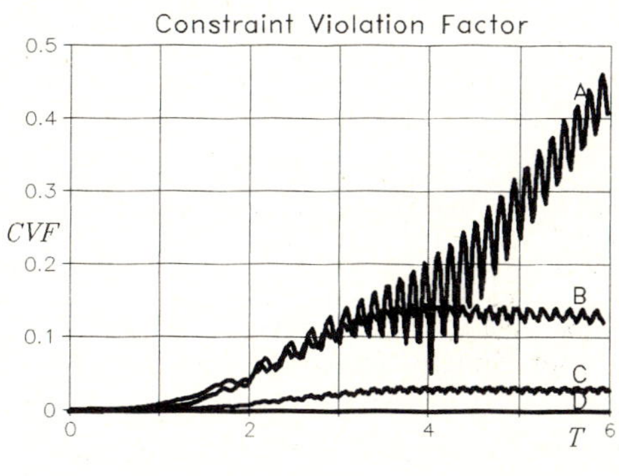

If Baumgarte's method is applied, the level of the constraint violation is stabilized (curve B), at least for the period during which there is no driving torque applied.

A time step reduction lessens at least in the same ratio the constraint violation factor, at the cost of a increase of the CPU time (curve C).

(Fig. 4.b)

5. Optimization of the Integration Scheme.

Among all the parameters used in the integration scheme, the step size Δt is undoubtedly the one on which the CPU time depends the most.
An adjustment of integration step size during the simulation is possible when variable step size algorithms as Newmark or Shampine-Gordon are used. The constraint violation factor provides the means to modify the step size Δt pertinently: in most cases, both the CPU time and the numerical errors can be reduced in this way.

Moreover, the Baumgarte's constraint violation stabilization technique has no reason to be applied when CVF doesn't reach a certain level. With a view to optimize the integration phase, an adjustment of the stabilization parameters a and b as a function of CVF is often interesting. (Cf. ref. [3]). In any case, the values of a and b used to damp out constraint error must be restricted to prevent affecting integration by considerable correction terms.

6. Constraint Violation Reduction Method.

When the geometric constraints are violated, we have:

$$CVF_g <> 0 \quad \text{et} \quad \{f_i(\{q_j\})\} = \{\zeta_i\} <> \{0\} \tag{Eq.6.a}$$

If a closed kinematic chain is characterized by M relative coordinates and F degrees of freedom, F coordinates can be arbitrarily chosen. The remaining N = M-F other parameters can be computed by using the loop equations: the geometrically assembled system verifies a null constraint violation factor. The choice of the N fixed coordinates among the M parameters remains ambiguous.

Nevertheless, it is possible to find the corrections for each of the M relative coordinates in such a way that:

- firstly, the loop-closure equations are fully satisfied at each time step: CVF = 0.
- secondly, the corrections are minimal: in other words, the corrected configuration is as close as possible to erroneous configuration. Moreover, all relative coordinates are treated in the same way.

It can be written that:

$$q_j = \bar{q}_j + \delta q_j \tag{Eq.6.b}$$

in which q_j are the consistent "exact" values, $\bar{q}_j$ the approximate values, δq_j , the corrections.

Let's write the following loop-closure equation:

$$T_1(\bar{q}_1 + \delta q_1).T_2(\bar{q}_2 + \delta q_2).....T_M(\bar{q}_M + \delta q_M).T_{M+1} = I \tag{Eq.6.c}$$

The first-order approximation of the transformation matrices hereabove will result in

$$T_j(\bar{q}_j + \delta q_j) = T_j(\bar{q}_j) + \left(\frac{\partial T_j}{\partial q_j}\right)_{\bar{q}_j} \delta q_j \tag{Eq.6.d}$$

The use of derivative operator matrices yields:

$$\frac{\partial T_p}{\partial q_j} = T_p \; . \; {}^p\Delta_j \tag{Eq.6.e}$$

This equation will give matrix products of the general form:

$$T_1(\bar{q}_1).T_2(\bar{q}_2).\cdots.T_M(\bar{q}_M).T_{M+1} \; +$$ (Eq.6.f)

$$T_1(\bar{q}_1).\Delta_1.\delta q_1.T_2(\bar{q}_2).\cdots.T_M(\bar{q}_M).T_{M+1} \; +$$

$$T_1(\bar{q}_1).T_2(\bar{q}_2).\Delta_2.\delta q_2.\cdots.T_M(\bar{q}_M).T_{M+1} \; +$$

$$\cdots \; +$$

$$T_1(\bar{q}_1).T_2(\bar{q}_2).\cdots.T_M(\bar{q}_M)\Delta_M.\delta q_M.T_{M+1} \; = I$$

The I^* matrix approaches the unit matrix:

$$(T_1(\bar{q}_1).\Delta_1.T_2(\bar{q}_2).\cdots.T_M(\bar{q}_M).T_{M+1}).\delta q_1 \; +$$ (Eq.6.g)

$$(T_1(\bar{q}_1).T_2(\bar{q}_2).\Delta_2.\cdots.T_M(\bar{q}_M).T_{M+1}).\delta q_2 \; +$$

$$\cdots \; +$$

$$(T_1(\bar{q}_1).T_2(\bar{q}_2).\cdots.T_M(\bar{q}_M)\Delta_M.T_{M+1} \;).\delta q_M = I \; - \; I^*$$

After development of each term, the following matrix equation is obtained:

$$\sum_{j=1}^{M} \begin{pmatrix} 0 & -\delta_z^j & \delta_y^j & d_x^j \\ \delta_z^j & 0 & -\delta_x^j & d_y^j \\ -\delta_y^j & \delta_x^j & 0 & d_z^j \\ 0 & 0 & 0 & 0 \end{pmatrix} \delta q_j \; = \; I - I^*$$ (Eq.6.h)

from which L independent algebraic equations in the M variables δq_j can be extracted. (Cf. ref. [4]).

For all the closed kinematic chains, we get the following indeterminate algebraic set of equations:

$$\left[C_{(\bar{q}_j)}\right]\{\delta q_j\} = \{\theta_j\}$$ (Eq.6.i)

in which the Jacobian matrix $[C_{(\bar{q}_j)}]_{N x M}$ can be recognized.

Instead of fixing arbitrarily M-N variables q_j, it will be ensured that the weighted sum of the corrections squared is minimal. We will realize the least-square condition:

$$\text{MIN} \; \sum_{j=1}^{M}(\omega_j.\delta q_j)^2$$ (Eq.6.j)

Weighting coefficients ω_j can be introduced in order to achieve dimensional homogeneity.

The solution of the problem can be easily found with recourse to Lagrange multipliers: the algebraic system to be solved can be written in matrix form as:

$$\begin{pmatrix} \begin{matrix} 2\omega_1^2 & 0 & \dots & 0 \\ 0 & 2\omega_2^2 & \dots & 0 \\ \dots & \dots & \dots & 0 \\ \dots & \dots & \dots & 0 \\ 0 & 0 & \dots & 2\omega_M^2 \end{matrix} & [C^T] \\ [C] & [0] \end{pmatrix} \begin{Bmatrix} \delta q_1 \\ \delta q_2 \\ \dots \\ \dots \\ \delta q_M \\ \lambda_1 \\ \lambda_2 \\ \dots \\ \dots \\ \lambda_N \end{Bmatrix} = \begin{Bmatrix} 0 \\ 0 \\ \dots \\ \dots \\ 0 \\ \theta_1 \\ \theta_2 \\ \dots \\ \dots \\ \theta_N \end{Bmatrix} \qquad \text{(Eq.6.k)}$$

The corrections $\{\delta q_j\}$ are easily found and the erroneous values $\{\overline{q}_j\}$ can be corrected.

The matrix of Eq.6.k has properties of sparsity and symmetry. The following algorithm is more computationnaly efficient. Firstly, the matrix expression Eq.6.l gives the Lagrange multipliers:

$$\left([C] \right) \begin{pmatrix} \frac{1}{2\omega_1^2} & 0 & \dots & 0 \\ 0 & \frac{1}{2\omega_2^2} & \dots & 0 \\ \dots & \dots & \dots & 0 \\ 0 & 0 & 0 & \frac{1}{2\omega_M^2} \end{pmatrix} \left([C]^T \right) \begin{Bmatrix} \lambda_i \end{Bmatrix} = - \begin{Bmatrix} \theta_i \end{Bmatrix} \qquad \text{(Eq.6.l)}$$

These ones are introduced in the next formula:

$$\delta q_j = \frac{\{-[C^T]\{\lambda\}\}_j}{2\omega_j^2} \qquad \text{(Eq.6.m)}$$

The method is iterative: a new set of position errors $\{\theta_i\}$ can be calculated and the Jacobian matrix must be reestimated from the new corrected values $\{q_j\}$. The process can be stopped as soon as CVF and/or $\sum_{j=1}^{M} (\omega_j . \delta q_j)^2$ are below user-fixed tolerances.

In practice, the convergence is realized extremely fast: this method will not affect the CPU time considerably.

7. Performance analysis of algorithms.

The crusher presented in fig. 4.a. only has one degree of freedom: it is possible to express one coordinate as a function of another: the relation between relative coordinates of joints 7 and 4 is presented in fig.7.e.

It is interesting to compare the plots (q_7 versus q_4) obtained for the two sets of integration parameters:

($a=b=0$, $\Delta t=0.025$s [A]), fig.4.c

($a=b=10$, $\Delta t=0.025$s [B]), fig.4.d.

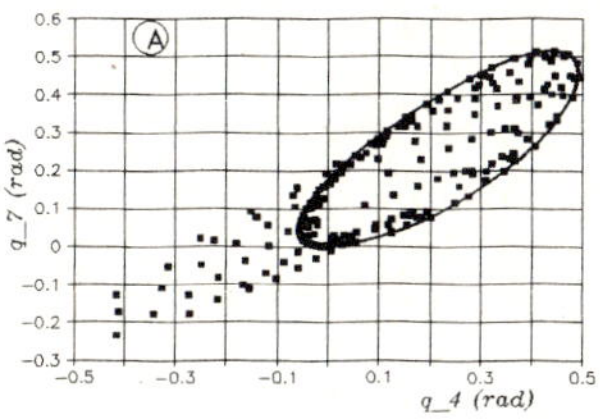

(Fig.7.c)

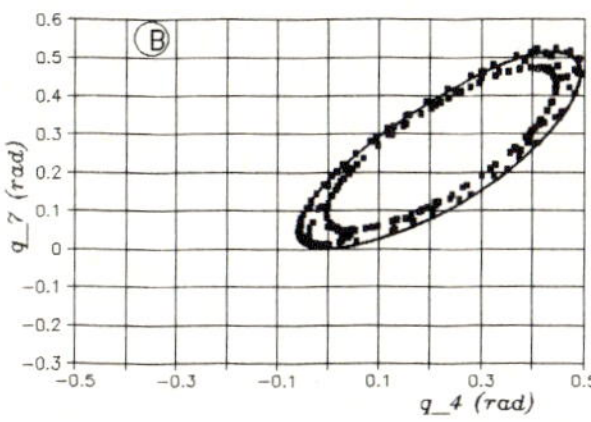

(Fig.7.d)

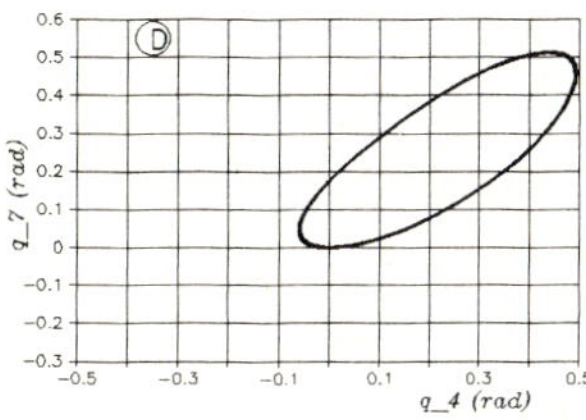

(Fig.7.e)

The Baumgarte's stabilization method brings convergence but it is unable to guarantee results verifying the geometric constraints. When used with a proper time step, this method yields accurate results.

If we only base our arguments on these curves, the Constraint Violation Reduction Method will appear the best. However, if must be emphasized that this algorithm always yields results consistent with the geometric constraints. A numerical error control based only on the CVF_g cannot be achieved anymore: other strategies are then available (kinematic constraints can be helpful).

Time evolutions of coordinate q_7 have been compared for integration achieved with a same time step but differing in the stabilization method: no one (fig. 7.f), Baumgarte's method with a = b = 10 (fig.7.g) and constraint violation reduction method (fig.7.h). The exact solution is shown in dotted lines. The last algorithm gives more stable and accurate results.

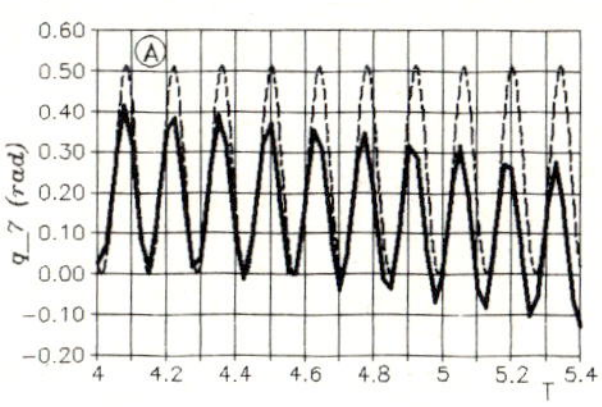

(Fig.7.f)

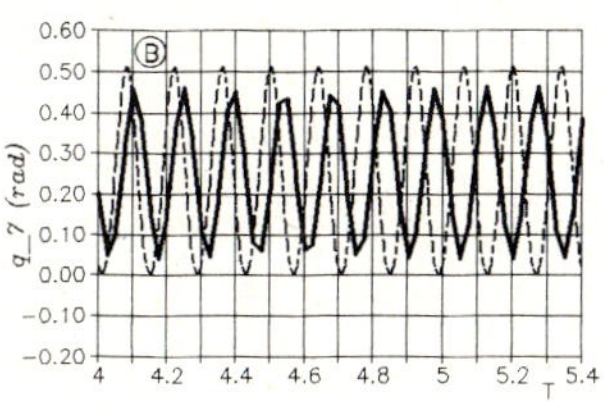

(Fig.7.g)

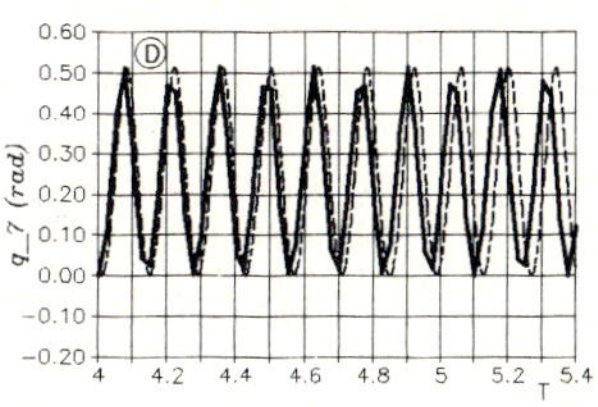

(Fig.7.h)

As shown in table 7.a, the computationnal efficiency of the Constraint Violation Reduction Method is satisfactory. The number of iterations needed to satisfy the convergence test in predictor-corrector scheme is less than in Baumgarte's method.

Method	Time step	Stabilization	Total number of iterations.	Maximal number of iterations by time step.
A	0.025	None a=b=0	2000	25
B	0.025	Baumgarte a=b=10	4748	76
C	0.005	Baumgarte a=b=10	6361	10
D	0.025	CVR-Method CVF==0	2840	57

(Tab.7.a)

8. Conclusions.

The definition of a Constraint Violation Factor allows to guarantee a global error below a user-fixed tolerance:

1° by an adaptative control of the parameters of the stabilization and integration scheme.

2° by a constraint violation reduction which consists of a judicious assembly of all the mechanism coordinates.

9. References.

[1] BAUMGARTE : Stabilization of Constraints and Integrals of Motion in Dynamical Systems. Computer Methods in Applied Mechanics and Engineering, Vol.1, 1972, pages 1-16.

[2] BECQUET : Validation théorique et expérimentale de modèles dynamiques direct et inverse de mécanismes articulés à structure cinématique complexe: application à la robotique. Thèse de Doctorat en Sciences Appliquées, Université Libre de Bruxelles, 1987.

[3] CHANG, NIKRAVESH : An Adaptative Constraint Violation Stabilization Method for Dynamic Analysis of Mechanical Systems. Transactions ASME, Journal of Mechanisms, Transmissions, and Automation in Design, Dec 85, Voo.107, pp 488-492.

[4] SHETH : A Digital Computer Based Simulation Procedure for Multiple Degree of Freedom Mechanical Systems With Geometric Constraints. Thèse University of Wisconsin, Ph.D., 1972

Stationary Motions of Mechanical Systems with an Infinite Number of Degrees of Freedom

V. G. DEMIN, YU. G. MARKOV, S. D. FURTA, N. I. CHURKINA

Moscow University and Moscow Aviation Institute

Summary

The report is devoted to the study of mathematical models of dynamics of mechanical systems with an infinite number of degrees of freedom, which has been recently carried out by a group of scientists headed by V. G. Demin. A contemporary approach to the study of motions of continuous systems is based on the functional analysis. In this case the motion equations are treated as ordinary differential equations in Banach spaces. The apparatus of differential equations in Banach spaces makes it possible to consider the problem on the basis of a maximally wide class of functions including generalized solutions of the problem stated. Bearing in mind the above, the authors have considered behaviour of different infinite-dimensional mechanical systems in the vicinity of stationary modes of motion.

The study can be divided into two parts. The first part is devoted to studying the evolution of motion of systems of rigid bodies with elastic and viscous-elastic elements. The main apparatus used by the authors in solving those problems is based on asymptotic methods of nonlinear mechanics. The basic theorems of the averaging method and the theory of singular perturbations for studying systems with an infinite number of degrees of freedom were used in works [1-4] .

The authors consider mechanical systems, the configurational space of which can be presented in the form $M \times \mathbb{H}$ where M is a smooth n-dimensional manifold with the local coordinates $\varphi=(\varphi_1,\dots,\varphi_n)$ which is diffeomorphic to the torus T^n; $\mathbb{H}$ is the Hilbert space of vector functions $\vec{u}$ describing the displacement of particles of the continuous medium. The motion of the deformed system as a whole is given via the canonical variables $I=(I_1,\dots,I_n)$, $\varphi=(\varphi_1,\dots,\varphi_n)$ on the cotangent fibration, and the motion equations can be presented in the following form

N. V. Banichuk · D. M. Klimov · W. Schiehlen (Eds.)
Dynamical Problems of Rigid-Elastic
Systems and Structures
IUTAM Symposium Moscow, USSR, May 23-27, 1990

$$\dot{I}_i = -\frac{\partial \mathcal{R}}{\partial \varphi_i}, \quad \dot{\varphi}_i = \frac{\partial \mathcal{R}}{\partial I_i}, \quad i = 1, \dots, n$$

$$\frac{d}{dt}\left(\nabla_{\dot{\vec{u}}} \mathcal{R}\right) - \nabla_{\vec{u}} \mathcal{R} + \vec{Q}_{\vec{u}} = 0 \tag{1}$$

Here $\mathcal{R}$ is the Routh functional of the problem, which is defined by the formula

$$\mathcal{R}[\vec{u}, \dot{\vec{u}}, \varphi, I] = \sum I_i \dot{\varphi}_i - T + \mathcal{E} + U$$

where $T = T[\vec{u}, \dot{\vec{u}}, \varphi, I]$ is the functional of the total kinetic energy of the system; $\mathcal{E} = \mathcal{E}[\vec{u}]$ is a quadratic form which is the functional of the potential energy of elastic deformations; $U = U[\vec{u}, \varphi]$ is the potential of the mass forces; $\vec{Q}_{\vec{u}}$ is a vector of the dissipative forces; ∇ is Frechet's differential. It is possible to use the asymptotic methods since in the system there exists a small parameter ε such that $\mathcal{R} = \mathcal{R}_0[I, \varphi] + O(\varepsilon^2)$. It is assumed also that $\vec{Q}_{\vec{u}} = O(\chi)$ where χ is another small parameter. When $\varepsilon = \chi = 0$, equations (1) correspond to the motion of the system as a rigid body. The main task is to study the evolution in the vicinity of stationary motions of the unperturbed system.

As an example, consider the motion of a viscous-elastic body with a rigid insertion in the central Newtonian field of forces. Assume that in an undeformed state the body is dynamically symmetric, the centre of mass of the system moves in an elliptical orbit, and the motion of the centre of mass does not depend on the motion about the centre of mass. The problem is solved on the basis of standard assumptions that Young's modulus of the elastic part of the system is large enough, and the time of damping of natural elastic vibrations due to the dissipation is much shorter than the characteristic time of the motion of the system as a whole.

The evolution process can be divided into two stages. At the first stage the vector of the moment of momentum $\vec{G}$ of the system will be either along the symmetry axis (when $C > A$) or in the equatorial plane of the inertia ellipsoid (when $A > C$) where A and C are equatorial and axial moments of inertia of the system in an undeformed state, respectively. At the second stage the axial rotation becomes slower and the angle between the vector $\vec{G}$ and the normal to the plane of the orbit tends to a certain stationary value.

The second part is devoted to the study of stability of equilibrium positions of systems with an infinite number of degrees of freedom. While solving these problems, one often faces a problem connected with the famous Lagrange-Dirichlet theorem of stability of equilibrium and its inversion. In the infinite-dimensional case the problem has its own spesific difficulties, viz. the stability in infinite-dimensional systems is studied with respect to certain integral characteristics of the norm type [5], an arbitrary choice of which can lead to different results of stability. On the other hand, these problems are characterized by all those difficulties found in the case of a finite number of degrees of freedom.

Here some general approaches to the study of stability of equilibria of infinite-dimensional systems with the help of Lyapunov's direct method will be considered.

Consider a mechanical system, the Lagrange function of which has the form

$$\mathcal{L}[\dot{u},u] = T[\dot{u},u] - U[u]$$

where $u \in \mathbb{H}_0, \dot{u} \in \mathbb{H}_1, \mathbb{H}_0 \hookrightarrow \mathbb{H}_1$, $\mathbb{H}_0, \mathbb{H}_1$ are some Banach spaces; $T[\dot{u},u]$ is the kinetic energy of the system, which is the positively defined quadratic form $\dot{u}$ when each $u \in \mathbb{H}_0$; $U[u]$ is the potential energy of the system. Assume that $u=0$ is an equilibrium position of the system, i. e. $\nabla_u U = 0$. Let the expansion of the potential energy U to the Maclaurin series begin with the quadratic terms

$$U = U_2 + \dots$$

If the system is finite-dimensional ($\dim \mathbb{H}_0, \mathbb{H}_1 < \infty$) and the form U_2 is positively defined, the equilibrium is stable. Analogous assertions have also been proved for the infinite-dimensional case [5-7] by means of analogues of the Lyapunov stability theorem with the total energy

$$\mathcal{H} = T + U \tag{2}$$

being considered as the Lyapunov function.

To study instability is more difficult. If the form can assume negative values, we can try to prove instability through the Chetaev functional

$$\mathcal{W} = (\nabla_{\dot{u}} T, u) \tag{3}$$

which is calculated on the paths with a negative store of energy. The functional (3) was suggested for finite-dimentional systems in [8]. The general instability theorem for the infinite-dimensional case where the functional (3) was used was proved in 5 . In [9,10] such functionals were applied to prove instability of concrete linear systems with an infinite number of degrees of freedom. Unfortunately, such a construction turns out to be of little use for most nonlinear systems, both finite-dimensional and infinite-dimensional ones. In [8] Chetaev suggested the following modification of the functional (3) for the finite-dimensional case

$$\mathcal{W} = (\nabla_{\dot{u}} T, f(u)) \tag{4}$$

where $f(u)$ was a certain vector field zeroing in the equilibrium. In [11] a concrete form of the vector field was found, which makes it possible to prove instability when the equilibrium is the only critical point of the first nontrivial form of the expansion of the potential energy. In [12] one of the authors suggested a similar construction of the vector field f in the infinite-dimensional case.

$$f = u - \varepsilon G \nabla_u U_2 \tag{5}$$

where $\varepsilon > 0$ is a certain small parameter; $G : \mathbb{H}_0^* \to \mathbb{H}_0$ is a smoothing operator. Note that the components of Chetaev's vector field f mean virtual displacements, therefore the smoothing is necessary for f to belong to $\mathbb{H}_0$.

The case when the equilibrium is not isolated is the most complicated. In this case there seems to be no general methods of constructing the vector field $\mathfrak{f}$. Below are the examples considered.

As one of the examples, the authors of the report consider the motion of a homogeneous elastic membrane in the central Newtonian field of forces, which is stretched over a rigid closed loop which is in circular Keplerian motion at a constant angular velocity with the centre situated in the centre of attraction. The loop is considered to be connected with a certain rigid body, the mass of which is large enough as compared with the mass of the loop and the membrane, and the centre of mass of which is in circular Keplerian motion, the body keeping its orientation relative to the orbital system of coordinates. A quasilinear partial differential equation of the second order is the motion equation of the system. The equation has a trivial equilibrium solution where the membrane is not deformed and its plane is perpendicular to the orbit. The authors have proved the following: if the membrane is situated inside a sphere with the radius equal to the radius of the orbit of the centre of mass of the body-carrier and the centre situated at the centre of attraction, then the above equilibrium is always stable; if the membrane is situated outside a sphere with the above radius, then the equilibrium becomes unstable when the values of stiffness are small enough. Here the stability is understood in terms of the norms of the Sobolev spaces. The stability has been proved via an analogue of the Lyapunov stability theorem with the total energy (2) being used as Lyapunov's functional; the instability has been proved by means of the Chetaev theorem with (3).

Then the planar motion of a nonlinear elastic thin rod (column) with the ends fastened with the help of hinges is considered. If the displacements of the ends of the column are given, there exists a rectilinear equilibrium form. The perturbed motion of such a system is described by two quasilinear hyperbolic partial differential equations with the polynomial non-linearity. In the problem reduced to the dimensionless form there exists an essential parameter. The stability of the rectilinear form of the column has been investigated for all values of that parameter except for the bifurcational ones. The stability with respect to the norms of the Sobolev spaces has been proved via an analogue of the Lyapunov stability theorem with the total energy (2) being used as Lyapunov's functional. The instability has been proved by means of the Chetaev theorem with (4) and (5) being used as Chetaev's functional and Chetaev's vector field.

Finally, the planar motion of a compound pendulum with an absolutely flexible unstrechable heavy thread attached to it is considered. The motion is described by two nonlinear integro-differential equations which are not solvable with respect to derivatives of higher order with respect to time. This system has an equilibrium solution where the thread is situated vertically below the point of suspension. Since the system of the motion equations is greatly nonlinear and is not solvable with respect to derivatives of higher order with respect to time, the problem of Lyapunov's stability of the given equilibrium solution is very complicated. The problem of stability with respect to two metrics can be solved in a much simpler way consisting in the synthesis of two concepts: Lyapunov's stability with respect to one metric and Lagrange's stability with respect to the other. Several different metrics are introduced on the phase space of the system, and stability is estimated with respect to the pairs of those metrics. The authors have proved the following. If the centre of mass of the system which is in equilibrium is below the point of suspension of the pendulum, the equilibrium is stable; if the centre of mass is above the suspension point, the equilibrium is unstable. The stability has been proved via an analogue of the Lyapunov stability theorem and the functional (2); the instability has been proved by means of the Chetaev theorem and the functional (4) with the vector field $\mathbf{f}$ having a rather complicated form. That can be explained by the following: in the vicinity of the physically feasible equilibrium position under consideration there exists the continuum of the so-called "folded" equilibrium positions where in some sections of the thread there appear negative tensions. In this respect the equilibrium position under consideration is not isolated, and the authors had to choose the components of Chetaev's field using only their intuition.

References

1. Klimov, D.M.; Markeev, A.P.: Dynamics of a nonhomogeneous elastic ring in the gravitational field. Moscow: Preprint No.331, Institute for Problems in Mechanics, the USSR Academy of Sciences, 1988.

2. Sidorenko, V.V.: Evolution of fast rotations of an elastic ring in the gravitational field. Moscow: Preprint No. 93, Institute of Applied Mathematics, the USSR Academy of Sciences, 1987.

3. Chernous'ko, F.L.; Shamaev, A.S.: Asymptotics of singular perturbations in the problem of dynamics of the rigid body with elastic and dissipative elements. Trans. of the USSR Academy of Sciences, Mechanics of Rigid Bodies No. 3 (1983), 33-42.

4. Vilke, V.G.: Analytical and qualitative methods of mechanics of systems with an infinite number of degrees of freedom. Moscow: Moscow University, 1986, 192.

5. Moiseev, N.N.; Rumyantsev, V.V.: Dynamics of bodies with cavities containing liquid. Moscow: Nauka, 1965, 439.

6. Movchan, A.A.: On stability of motion of continuous bodies. The Lagrange theorem and its inversion. Eng. col. Vol. 29 (1968), 3-20.

7. Sirazetdinov, T.K.: Stability of systems with distributed parameters. Novosibirsk: Nauka, Sib. Dept. 1987, 232.

8. Chetaev, N.G.: On stability of equilibrium in some cases when the function of forces is not the maximum. Appl. Mathem. and Mech. Vol. 16 (1952) No. 1, 89-93.

9. Movchan, A.A.: On Lyapunov's direct method in problems of stability of elastic systems. Appl. Mathem. and Mech. Vol. 23 (1959) No. 3, 489-493.

10. Vladimirov, V.A.: On instability of equilibrium of nonhomogeneous liquid when potential energy is not minimal. Appl. Mathem. and Mech. Vol. 52 (1988) No. 3, 415-422.

11. Palamodov, V.P.: On stability of equilibrium in the potential field. Functional analysis and its application, Vol. 11 (1977), No. 4, 42-55.

12. Furta, S.D.: On the loss of equilibrium of a flexible unstretchable thin rod with one fixed end in the gravitational field. The Moscow University Bulletin, ser. 1, mathem, mech., No. 6, 37-42.

A Hierarchical Optimization Procedure for the Modelling of Mechanical Systems

Eschenauer, H. A.; Müller-Slany, H. H.
Research Laboratory for Applied Structural Optimization
Institute of Mechanics and Control Engineering
University of Siegen, D-5900 Siegen, Germany

Abstract: Substantially condensed lumped-mass systems are applied to simulation studies of complex elasto-mechanical systems. In case of a heuristic discretization, such models show unacceptably high errors in the dynamic system properties. It is shown how dynamically precise simple lumped-mass systems can be generated by means of a hierarchical optimization procedure. The model building of a complex three-dimensional truss structure of a parabolic antenna reflector (10m-SMT) has been chosen as an example.

1. The modelling problem

Analysis and simulation problems of complex elasto-mechanical systems require simple physical substitute models with reduced degrees of freedom in the form of lumped mass or multi-body systems. The most important demands on those models are:

- *Consistency* with the *dynamic properties* of the complex real system,
- *Consistency* with the *experimental data* for those systems which are difficult to describe in mathematical terms,
- *Consistency* with *desired dynamic properties* in the sense of an optimal system design.

The dynamic properties which occur in dependence on the simulation purpose are the theoretical, measured, and planned parameters describing the motion behaviour of an elasto-mechanical system. These are the *mass-geometric values* (total mass, position of the centre of mass, elements of the inertia tensor), *modal values* (the eigenfrequencies of interest and eigenmodes) as well as *linear momentum, angular momentum,* and *kinetic energy* of the system in the state of eigenvibration.

Simplified physical models are generally established by means of heuristic methods carrying out an arbitrary discretization of the continuum. Due to a strong reduction of order, these models show large errors within the dynamic properties. They are unsuitable for the subsequent simulation tasks, as e.g. the layout of control systems or the determination of efficiency limits of the planned real system.

The heuristic modelling mentioned above shall now be replaced by an *Adapted Model.* The latter is a physical model the dynamic properties of which are adapted to the properties of the real system by means of optimization procedures. This accurate model of a low order is very suitable for various simulation tasks. The

N. V. Banichuk · D. M. Klimov · W. Schiehlen (Eds.)
Dynamical Problems of Rigid-Elastic Systems and Structures
IUTAM Symposium Moscow, USSR, May 23-27, 1990

selection of the real system properties to be adapted depends on the simulation objective for which the adapted model is built up. Fundamental for this task are the *mass-geometric properties*: total mass, position of the centre of mass, and elements of the inertia tensor. The correct projection of these parameters into the adapted model determines the rigid-body-motion behaviour and the realistic values of the stress resultant forces and of control forces in control engineering systems. The eigenfrequencies and eigenvectors in the frequency range under investigation are the basis of the mathematical discretization and condensation of system components when using adapted models in the substructural analysis and synthesis for tree-structures [1]. These *modal properties* are the second important range of adaptation. It can be shown, however, that a substructural synthesis from adapted models does not provide an accurate projection of the complete system if *only* properties from these two ranges of adaptation are fulfilled. The demand for the adaptation of linear momentum and angular momentum vectors of the substructure in the state of the investigated eigenvibrations gains an important influence on the result of the substructural synthesis.

The problem of model adapttion, i.e. the reconstruction of the mass matrix and the stiffness matrix with the given spectral characteristics of the system, has been dealt with in various ways in the literature: by methods of inverse eigenvalue problems [2], [3] or by *Modal Pertubation Methods* in the range of the *Redesign-Problem* [4]. So far, an adaptation of the model to the linear momentum and angular momentum properties of the original could not be achieved by either way. The model adaptation can be formulated as a constrained vector optimization problem as occuring in the case of structural optimization problems. Parallel to that, a great number of single objective functions which describe the error for a system property between the original and the model, have to be minimized. The representation of this task by means of a scalar optimization problem cannot use the strategy applied to structural optimization problems but is to follow a new one: the *hierarchical optimization procedure*.

The basic idea of this method is the gradual adaption of the model to different system properties. This yields levelled optimization problems which can be solved one by one. The properties already adapted in previous stages are fixed in the following optimization steps by means of additional equality constraints. This procedure requires an optimization algorithm which adapts to highly non-linear optimization problems and which works in the feasible range in every main iteration in order to secure already adapted system properties. Details on the optimization procedure can be found in [5, 6].

2. Formulation of the model adaptation as an optimization problem

The generation of a simple physical model as a lumped-mass system or a multi-body system requires the solution of two problems:

1. The designing of a suitable model structure for the solution of the given simulation task.
2. The search for the system parameters in order to minimize the error between given dynamic properties of the model and of the original system.

In order to solve the first problem, the experience of the engineer as the decision maker is needed. Formal solution methods cannot necessarily be applied here. The solution of the second problem can be formulated as a non-linear, constrained vector optimization problem, to be solved by direct methods from the mathematical programming. In an iterative programme-controlled process, the parameters of the model to be adapted are varied until the given characteristic properties are equal to the values of the original system, taking into account a residual error.

The following assumptions have to be considered for the Initial Design of the reduced model as a lumped-mass system:

- The kinematics of a model with a reduced number of degrees of freedom have to be determined according to the projected eigenmodes.
- The model ought to describe important points in the system, e.g. sensor points.
- Interface sections to structural elements have to be modelled closely to reality.
- Free vibrations of the model must not contain eigenmodes which the original system cannot carry out.

The parameters of this model are the design variables $\mathbf{x} \in \mathbb{R}^n$ in the optimization problem, e.g. geometric values (coordinates, cross-sectional areas, moments of inertia) and masses. For a given problem, the task of the model adaptation can be divided into three sub-problems:

1. Determination of those system properties which are significant for the modelling objective; Setting of the highest eigenfrequency still in the range of interest for the modal reduction.
2. Designing of a physical model and formulation of the model adaptation as a constrained vector optimization problem.
3. Development of a solution strategy for the vector optimization problem by the introduction of scalar substitute objective functions: designing of the *hierarchical optimization procedure.*

The adapted model is to possess given dynamic properties (according to Table 1), depending on the simulation objective. The realization of each particular system property $E_i(\mathbf{x})$ as a scalar quantity or $\mathbf{E}_i(\mathbf{x})$ as a vector quantity, respectively, will be formulated as a single objective $f_i(\mathbf{x})$ and will be described by an error functional, depending on the vector $\mathbf{x}$ of the design variables. All $i = 1, \ldots, m$

error functionals together form the components of the objective function vector $\mathbf{f(x)} \in \mathbb{R}^m$. For scalar and vectorial dynamic properties the following expressions are valid:

$$f_i(x) = 1 - \frac{E_i^M(\mathbf{x})}{E_i^O} \; ; \; f_i(\mathbf{x}) = \frac{|\mathbf{E}_i^O - \mathbf{E}_i^M(\mathbf{x})|}{|\mathbf{E}_i^O|} \; ; \; i = 1, \dots, k \tag{1}$$

with $\mathbf{x}$ vector of the design variables; E_i, $\mathbf{E}_i$ system property according to Table 1, O original, M model.

criterion	Lumped-Mass-System with p points of mass	influence on dynamics
	1. Mass-geometric properties	
total mass	$M = \sum_{i=1}^{p} m_i$	rigid-body motion, global forces and moments in the system
centre of mass	$\underline{r}_{OS} = [r_x, r_y, r_z]^T$	
inertia tensor	$\underline{\underline{\Theta}}_S = \begin{bmatrix} \Theta_{xx} & \Theta_{xy} & \Theta_{xz} \\ \Theta_{xy} & \Theta_{yy} & \Theta_{yz} \\ \Theta_{xz} & \Theta_{yz} & \Theta_{zz} \end{bmatrix}$	
	2. Elasto-static properties	
deformation behaviour	$\underline{r} = \underline{\underline{K}}^{-1} \underline{F}$	conditions for correct eigenmodes
	$\underline{r}_j = \underline{\underline{K}}^{-1} \mathrm{diag}[m_1, \dots, m_p] \, \omega_j^2 \underline{q}_j$	
	3. Modal properties	
eigen-frequencies	ω_j , $j = 1, \dots, k$	correct eigenmodes
eigenmode	$\underline{q}_j$, $j = 1, \dots, k$	
linear momentum	$\underline{p}_j = \sum_{i=1}^{p} m_i \omega_j \underline{q}_j^i$, $j = 1, \dots, k$	correct stress resultant forces in synthesis problems
angular momentum	$\underline{L}_{Oj} = \sum_{i=1}^{p} \underline{r}_{Oi} \times \underline{p}_j$, $j = 1, \dots, k$	
kinetic energy	$T_j = \omega_j^2 \underline{q}_j^T \mathrm{diag}[m_1, \dots, m_p] \underline{q}_j$ $j = 1, \dots, k$	vibration energy of the system

Table 1. Dynamic properties of the adapted model

The errors f can be minimized by varying the design variable $\mathbf{x}$, i.e. the model is adapted to the property of the original system which makes it possible to formulate the model adaptation as a constrained vector optimization problem:

$$\min_{\mathbf{x} \in \mathbb{R}^n} \{ \mathbf{f(x)} | \, \mathbf{h(x)} = \mathbf{0} \, ; \, \mathbf{g(x)} \le \mathbf{0} \} \tag{2}$$

with $\mathbf{f}(\mathbf{x})$: objective function vector,
$\mathbf{x}$: vector of the design variables,
$\mathbf{g}(\mathbf{x})$: vector of the p inequality constraints,
$\mathbf{h}(\mathbf{x})$: vector of the q equality constraints.

We now have to find vector $\mathbf{x}_{opt} \in X$ which leads the objective function to take on a minimum, fulfilling the constraints. The feasible design space is limited by bounds:

$$X := \{ \mathbf{x} \in \mathbb{R}^n \mid \mathbf{x}_l \leq \mathbf{x} \leq \mathbf{x}_u \} \quad . \tag{3}$$

As there are several objective functions which have to be minimized at the same time, no design variable vector can be given which satisfies all demands optimally in the case of conflicting objectives. The solution vectors of the minima of the various. objective-function components differ. There are different ways of forming a compromise solution. All methods reduce the vector optimization problem to a scalar optimization problem. In case of the model adaptation, the *minimization of a preference function* is relevant: A compromise solution is achieved by the scalar minimization of a preference function or a substitute objective function which result from the superposition of the various objective functions:

$$p[\mathbf{f}(\mathbf{x})] = \sqrt{\sum_{i=1}^{m} w_i \, [f_i(\mathbf{x})]^2} \, , \tag{4}$$

with: $p[\mathbf{f}(\mathbf{x})]$: preference function, w_i: weighting factors.

3. The hierarchical optimization procedure

The hierarchical optimization uses a strategy of successive minimization of simplified preference functions: In this method, the *final design* of one optimization step serves as the *initial design* of the following stage in which already minimized cost functionals are used as further constraints. The hierarchical optimization procedure is controlled by setting the weighting factors $w_i = 1$ or $w_i = 0$. By this, the weighting factors were use as control coefficients. Until the adapted model is achieved, the model building in the hierarchical optimization is structured by varying the numbers w_i in several successive steps. Each step is especially characterized by adding further equality constraints which fix the already adapted properties:

***1st optimization step*:** The objective function contains the errors of the mass-geometric quantities: centre of mass and the elements of the inertia tensor. The correct total mass is given in the *initial design* and is fixed by means of an equality constraint

***2nd optimization step*:** The mass-geometric system adaptation determines the rigid-body properties of the model. For given complex elasto-mechanical systems the elasto-static deformation behaviour of the system is chosen in such a way that the static force system from the maximum inertia forces in an eigenvibration displaces the elasto-mechanical structure precisely into the shape of the corresponding eigenmode. The mass-geometric properties remain fixed in this stage.

3rd **optimization step**: The *initial design* for the third optimization step is generated by means of the adapted elasto-mechanical properties of the FE-structural elements. The mass-geometric quantities are fixed by constraints and the objective function contains the error of the eigenfrequencies ω_i.

4th **optimization step**: In addition, the adapted eigenfrequencies are introduced as constraints. The objective function contains the errors of the eigenmodes $\mathbf{q}_i$, and of the vectors of linear momentum $\mathbf{p}_i$ and of angular momentum $\mathbf{L}_{0i}$.

The result of the fourth optimization step is the adapted model as the *final design.*

4. Example: An Adapted Model for the reflector of a radio telescope

The aim of the modelling process is a simple lumped-mass model for the reflector of the 10-m-SMT (submillimeter wave telescope with a reflector diameter of 10 m [7]). This model should represent the mass-specific data (mass, position of the centre of mass, moments of inertia) and the frequencies of the first eight eigenvibrations with high accuracy. The eigenmodes should be qualitatively precise and in the correct sequence. The reflector is a truss structure made of bars, beams, and membrane elements (Fig. 1) covered with panels. The FE-model of the structure shows more than 900 degrees of freedom. The dynamic properties (eigenfrequencies and eigenmodes) are compiled in Table 2 as a result of computations of the reflector with a complete FE-model.

Mode i	Frequency ω(Hz)	Vibration mode
1	22.88	torsional vibration around axis of symmetry
2 3	37.03 37.03	bending vibration around lateral axis (double eigenvalue, two different modes)
4 5	42.18 42.33	shell vibration with 4 nodal lines (different eigenvalues and modes)
6 7	50.30 50.30	shell vibration with 6 nodal lines (double eigenvalue, 2 diff. modes)
8	55.56	bending vibration in direction of axis of symmetry

Table 2. Vibration modes 1 to 8 of the reflector

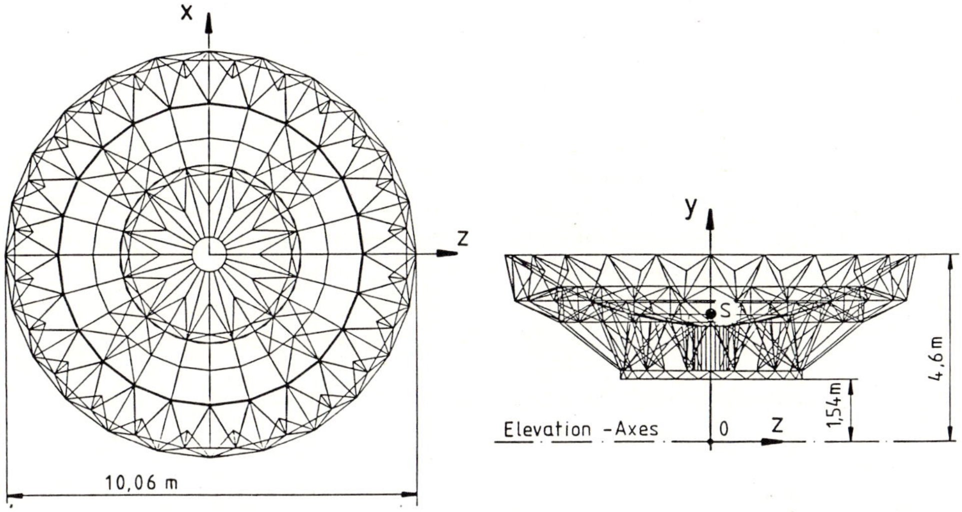

Fig. 1. Reflector of a parabolic antenna (10m-SMT)
Reflector values:
mass: m = 4627 kg ; centre of mass: y_S = 3050 mm
moments of inertia: $\Theta_{xxO} = \Theta_{zzO} = 68180$ kg m^2; $\Theta_{yyO} = 42760$ kg m^2.

According to Fig. 2, the beam structure with discrete mass points is designed as an initial design. The FE-model consists of 40 beam elements with 144 degrees of freedom. All elements can be correlated with seven different beam groups taking into account that according to the real structure the four diagonal ring-lacer are stiffer than the other four. This constructive design leads to slightly different frequencies of the 4th and 5th eigenvibration.

In order to determine design variables, necessary constructive connection points have to be established first, like the diameter of the inner ring and the additional points on the ring-lacer. The diameter of the inner ring describes the connection points of the reflector at the elevation frame, and the points on the ring-lacer refer to the couple points of the subreflector construction. All other system values have to be adapted until the final design corresponds to the objectives mentioned before. As the model, with the exception of the ring-lacer, is axially symmetric, the system variables can be combined in groups.

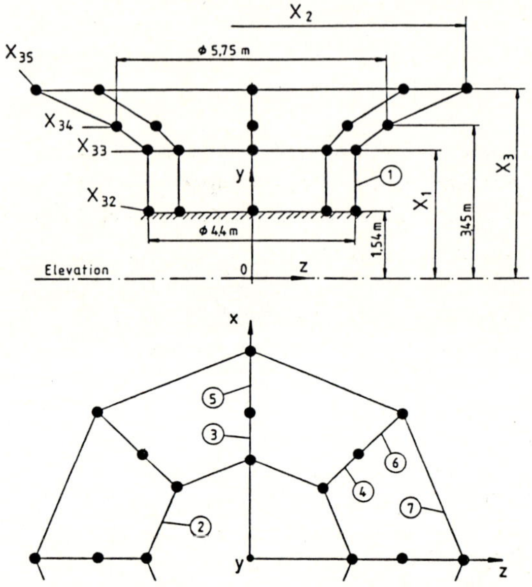

Fig. 2. Lumped-mass system for the reflector
$x_1 \dots x_{35}$: design variables, 1 ... 7 : beam elements

Thus, an optimization problem with n = 35 design variables is obtained:

$\mathbf{x} = [\, x_1,\dots,x_{35}]^T$:

$x_{1\dots3}$	3 position coordinates,
$x_{4\dots10}$	7 beam cross-sections,
$x_{11\dots31}$	21 moments of inertia of the beams,
$x_{32\dots35}$	4 mass points.

The optimization problem for adapting the model to the modal data and to the mass-specific parameters of the real system yields the vector optimization problem according to Table 3.

In order to scalarize the vector optimization problem of Table 3, the following preference function is introduced:

$$p[\mathbf{f}(\mathbf{x})] = \sum_{k=1}^{5} w_k \; f_k(\mathbf{x}). \tag{5}$$

The optimization hierarchy is controlled by the weighting factors $w_k = 1$ or $w_k = 0$, $k = 1, \dots, 5$. In the corresponding scalar optimization problem

$$\min_{\mathbf{x} \in X} \{p[\mathbf{f}(\mathbf{x})] : \mathbf{h}(\mathbf{x}) = \mathbf{0}\}, \quad \mathbf{x} = [x_1, \ldots, x_{35}]^T, \tag{6}$$

$$X := \{ \mathbf{x} \in \mathbb{R}^n : \mathbf{x}_l \leq \mathbf{x} \leq \mathbf{x}_u \}$$

equality constraints according to Table 4 are introduced in order to fix already adapted properties. The residual errors of the lumped-mass system are taken into consideration by the parameters ε_i in each adaptation stage. Table 4 presents the complete hierarchical optimization procedure.

$$\min_{\underline{x} \in X} \{\underline{f}(\underline{x})\}, \quad \underline{x} = [x_1, \ldots, x_n]^T, \quad n = 35$$

$$X := \{\underline{x} \in \mathbb{R}^n \,|\, \underline{x}_l \leq \underline{x} \leq \underline{x}_u\}$$

$$\underline{f}(\underline{x}) = \begin{bmatrix} \left(1 - \frac{m^M}{m^O}\right)^2 \\ \left(\frac{y_{os}^O - y_{os}^M}{y_{os}^O}\right)^2 \\ \sum_{x,y} \left(1 - \frac{\Theta_{ii}^M}{\Theta_{ii}^O}\right)^2 \\ \sum_{j=1}^{8} \left(\underline{q}_j^O - \underline{r}_j^M\right)^2 \\ \sum_{j=1}^{8} \left(1 - \frac{\omega_j^M}{\omega_j^O}\right)^2 \end{bmatrix}$$

M: Modell

O: Original

$\|\underline{q}_j\| = \|\underline{r}_j\| = 1$

$\underline{q}_j$ = eigenvector

$\underline{r}_j$ = displacement vector in case of loading by inertia forces of the j-th eigenmode

Table 3. Optimization problem for the adaptation process

The *final design* of the optimization procedure is an highly accurate *adapted model* with adapted mass properties and adapted eigenfrequencies $\omega_1 \ldots \omega_8$. The residual error is in the range of $\varepsilon = 10^{-8}$. The correct mode sequence could be secured by an Initial Design interactively developed at the computer on the basis of the results of the second optimization step (Table 4). Figs. 3 and 4 show the eigenmodes $\mathbf{q}_6(\mathbf{x})$ and $\mathbf{q}_8(\mathbf{x})$ of the real system and of the adapted model.

Initial design

1st optimization step		weighting factors
preference function	$p[\underline{f}(\underline{x})] = \left[\left(1-\frac{y^M_{OS}}{y^O_{OS}}\right)^2+\left(1-\frac{\Theta^M_{xx}}{\Theta^O_{xx}}\right)^2+\left(1-\frac{\Theta^M_{zz}}{\Theta^O_{zz}}\right)^2\right]$	$w_{2,3} = 1$ $w_{1,4,5} = 0$
equality constraints	$h_1(\underline{x}) = 1 - \frac{m^M}{m^O} - \varepsilon_1 = 0$	

2nd optimization step		weighting factors
preference function	$p[\underline{f}(\underline{x})] = \sum_{j=1}^{8}\left(\underline{q}_j^O - \underline{r}_j^M\right)^2$	$w_4 = 1$ $w_{1,2,3,5} = 0$
equality constraints	$h_1(\underline{x}) = 1 - \frac{m^M}{m^O} - \varepsilon_1 = 0$ $h_2(\underline{x}) = 1 - \frac{y^M_{OS}}{y^O_{OS}} - \varepsilon_2 = 0$ $h_3(\underline{x}) = 1 - \frac{\Theta^M_{xx}}{\Theta^O_{xx}} - \varepsilon_3 = 0$ $h_4(\underline{x}) = 1 - \frac{\Theta^M_{yy}}{\Theta^O_{yy}} - \varepsilon_4 = 0$	

3rd optimization step		weighting factors
preference function	$p[\underline{f}(\underline{x})] = \sum_{j=1}^{8}\left(1-\frac{\omega_i^M}{\omega_i^O}\right)^2$	$w_5 = 1$ $w_{1,\ldots,4} = 0$
equality constraints	$h_1(\underline{x}) = 1 - \frac{m^M}{m^O} - \varepsilon_1 = 0$ $\vdots$ $h_4(\underline{x}) = 1 - \frac{\Theta^M_{yy}}{\Theta^O_{yy}} - \varepsilon_4 = 0$	

Final design: Adapted Model

Table 4. Hierarchical optimization procedure

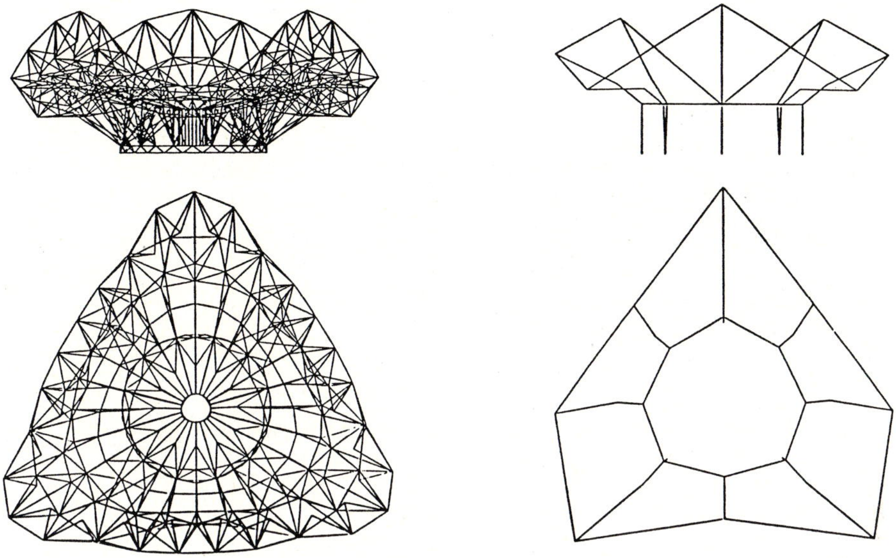

Fig. 3. 6th eigenmode: ω_6 = 50.30 Hz

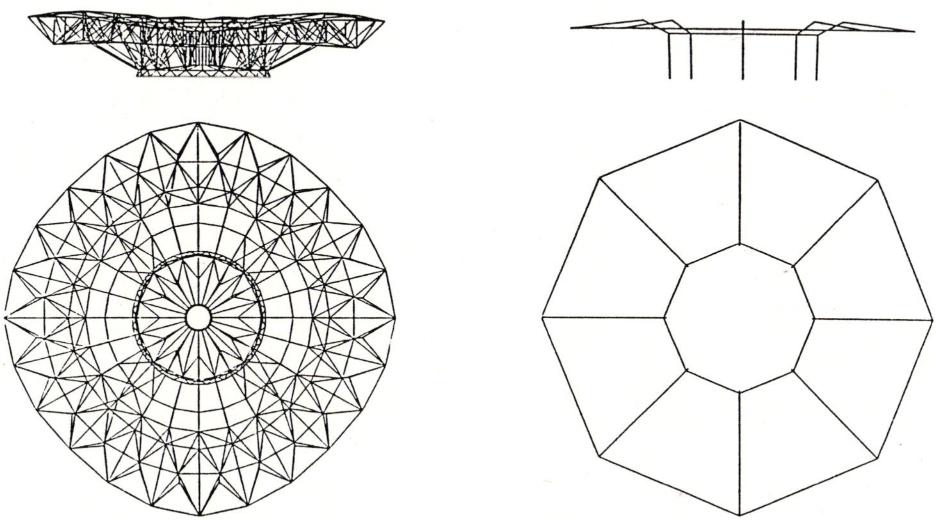

Fig. 4. 8th eigenmode: ω_8 = 55.56 Hz

5. Conclusion

The generation of simple lumped-mass systems with adapted dynamic properties can be formulated as a constraint vector optimization problem. Efficient optimization methods are developed for problems in the field of structure optimization. These methods can be used successfully, when the problem of model adaptation is formulated as a hierarchical optimization procedure. This is shown for generating a simple lumped-mass system for the reflector structure of a 10-m-submillimeter radio telescope.

References

[1] MÜLLER-SLANY, H.H.: Substruktursynthese für Schwingungsuntersuchungen mit angepaßten Modellen. ZAMM, Vol. 71 (to appear 1991).

[2] BIEGLER-KÖNIG, F.: Inverse Eigenwertprobleme. Dissertation. Universität Bielefeld, 1980.

[3] GLADWELL, G.M.L.: Inverse Problems in Vibration. Martinus Nijhoff Publishers, Dordrecht, 1986.

[4] SANDSTRÖM, R.E./ANDERSON, W.J.: Modal Perturbation Methods for Marine Structures. SNAME Transactions, Vol. 90, 1982, p 41-54.

[5] BREMICKER, M.: Dekompositionsstrategie in Anwendung auf Probleme der Gestaltsoptimierung. Dissertation, Universität Siegen. Fortschritt-Berichte VDI, Reihe 1, Nr. 173. VDI-Verlag, Düsseldorf, 1989.

[6] ESCHENAUER/KOSKI/OSYCZKA: Multicriteria Design Optimization - Procedures and Applications. Springer Verlag, Berlin/Heidelberg/New York, 1990.

[7] MÄDER/STENVERS/GATZLAFF/WILMS: Das 10-m-Submillimeterwellen-Teleskop. Techn. Mitt. Krupp 1/1990, S. 27 - 42.

Mechanics of Multi-body Systems

A.Yu.ISHLINSKY

Institute for Problems in Mechanics
U S S R Academy of Sciences
Moscow, U S S R

Summary

This paper gives a survey of basic scientific directions in mechanics of a rigid body. In the middle of this century the new branch of mechanics, the mechanics of systems of rigid bodies, has appeared and now it is succesfully developing. This is especially valid for the theory of rotation of a rigid body on a string which is a limit case for the motion of a couple of bodies. The great variety of modes of steady-state rotation of a body on a string were found. Some of these motions look like paradoxes. The investigation of stability of steady-state motions and their bifurcations often leads to considerable mathematical difficulties.
Dynamic balancing of bodies and the design of large centrifugal machines give graphic examples of practical applicability of this new branch of mechanics.
The motion of a rigid body with a cavity partially or completely filled with an ideal or viscous fluid is closely connected to the dynamics of a rigid body with a fixed point or hung on a string. In this field new important phenomena and engineering applications were also found.

Our Symposium has already a definite history. Its first assembly was held in Munich about ten or even more years ago. It was organized by IUTAM on the initiative of a remarkable scientist Kurt Magnus. The name of Prof. K.Magnus has been already mentioned today in greeting addresses. I am pleased to note that the books of the Professor were translated into Russian, with his agreement, and had a great success in this country. Regrettably, Prof.K.Magnus could not travel to Moscow for the Symposium. I think it will be good if we send him a greeting telegram on behalf of the Symposium. I will return again to the name of K.Magnus in connection with one important formula of the theory of a gyroscope.

N.V. Banichuk · D. M. Klimov · W. Schiehlen (Eds.)
Dynamical Problems of Rigid-Elastic
Systems and Structures
IUTAM Symposium Moscow, USSR, May 23-27, 1990

The scope of our Symposium gradually changes. In the beginning it was only mechanics of rigid bodies. Now it already includes mechanics of rigid and elastic systems and structures.

I believe that new branches of mechanics of multi-body systems will occupy greater room in agendas of future Symposia, e.g. servo systems, inertial navigation, robotics, walking machines, etc. By the way, I see here the scientists who just deal with the last problem. They are Professor V.V.Beletsky and Professor E.A.Devyanin.

Now I would like to add some words concernig the mechanics of systems of rigid bodies itself. It may look rather astonishing, but this branch has only a half-century history. I remember that about I940 I met L.G.Loytsansky and A.I.Lurje,the authors of the well-known course of theoretical mechanics. We mentioned that in textbooks on theoretical mechanics (sometimes we call it rational mechanics) there are the following sections: mechanics of a particle, mechanics of systems of particles, mechanics of a rigid body. But mechanics of systems of rigid bodies is absent.

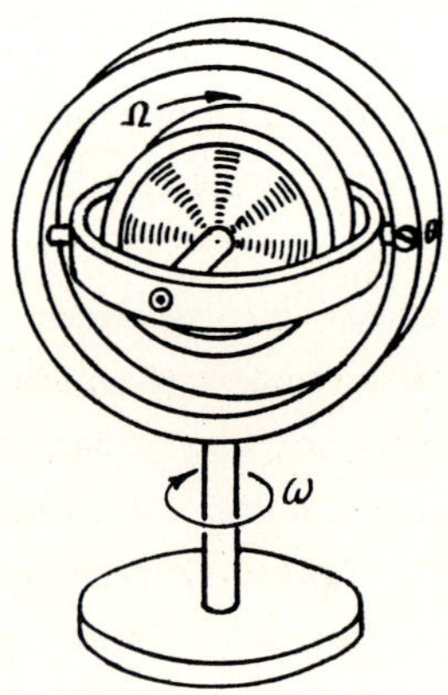

Fig.1

One of the examples of mechanics of systems of rigid bodies is the dynamics of a gyroscope in gimbals. Only in I939 our scientist E.L.Nikolai [1] derived the equations for this mechanical system, and N.G.Chetayev [2] undertook their integration in hypergeometric functions. They noticed that a free gyroscope in gimbals, frictionless but having had been vibrating, could change the average position of its axis, i.e. deviate. I will not show you the gyroscope in gimbals. You all know it very well (Fig.1). K.Magnus [3] was the first to suggest (in 1955) the formula for the calculation of such deviation. It is the only formula I intend to give in this paper. Here it is:

$$\left\langle \frac{d\alpha}{dt} \right\rangle = - \frac{\dot{\alpha}_a^2 (A_2 + C_1) \, tg\beta_0}{2H \cos\beta_0} .$$

Here β_0 is the average deviation of the axis of the gyroscope's rotor from the position, in which the axis is perpendicular to the plane of the inner ring; α_a is the amplitute of vibrations of the outer ring, and $\langle d\alpha/dt \rangle$ is the average angular velocity of the deviation.

For many specialists in mechanics of gyroscopes this formula was like a bolt from the blue. The gyroscope without friction does not conserve its orientation with respect to the fixed star, unlike all have been used! Sometimes hot discussions had arisen. It is interesting that B.T.Plymale and R.Goodstein [4] obtained just the same formula and almost at the same time, but, however, somewhat later than K.Magnus. This formula is approximate, since it was derived using the method of averaging. And great efforts were undertaken to estimate its accuracy. I am pleased to note that the Director of our Institute Prof.D.M.Klimov and N.P.Stepanenko, who is present at this Symposium, could integrate the equations of this problem in closed form for a special relationship between moments of inertia [5]. By this they answered the question, to what extent the Magnus formula was justified. Subsequently Ya.L.Luntz [6], A.D.Bryuno [7], K.G.Valeev [8], the scientist from our Institute V.F.Zhuravlyov [9] and the author of this paper [10] made estimates and improvements for this formula in general case.

The second example of mechanics of systems of rigid bodies is also well-known for a long time and I even find it difficult to call the scientist who was the first to deal with it. It is the

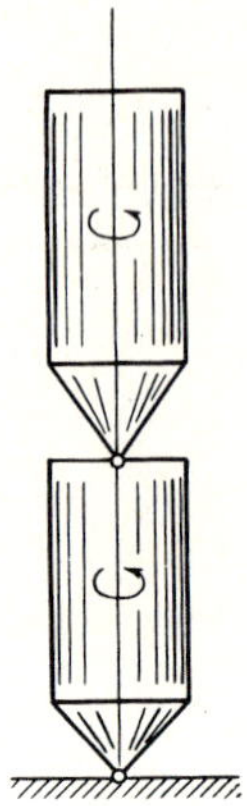

Fig.2

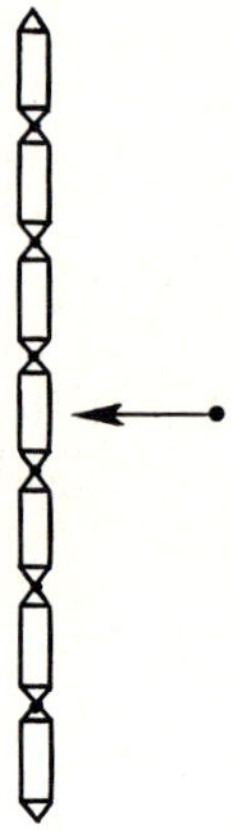

Fig.3

stability problem for two tops of Lagrange placed one on the other (Fig.2).In particular, P.V.Kharlamov considered the chain of tops of Lagrange connected with each other [11]. Professor W.Schiehlen, the guest of our Institute, treated an interesting example of motion of a chain of rigid rods connected by hinges (Fig.3). Prof. W.Schiehlen calculated on a computer the motion of all the rods after one of them had been struck [12]. He presented the paper devoted to this topic for the first session of our Symposium.

One can tell about many other interesting investigations refering to the problem of motion of systems of rigid bodies. Here is an example which was treated in different periods by three women-specialists in theoretical mechanics. They considered two tops connected by a spherical hinge (Fig.4). If the tops are forced to vibrate, the system will rotate around the bisector of the angle between their axes. It was discovered by N.P.Stepanenko [13]. L.N.Rogachova took part in solving this problem for the case of different tops [14]. M.E.Lesina, the pupil of P.V.Kharlamov, added the elasticity [15]. Of course, all such motions take place only in the absence of gravity.

Now I would like to give you an example with double pendulum, the history of which goes back to Huygens. Imagine a beam and

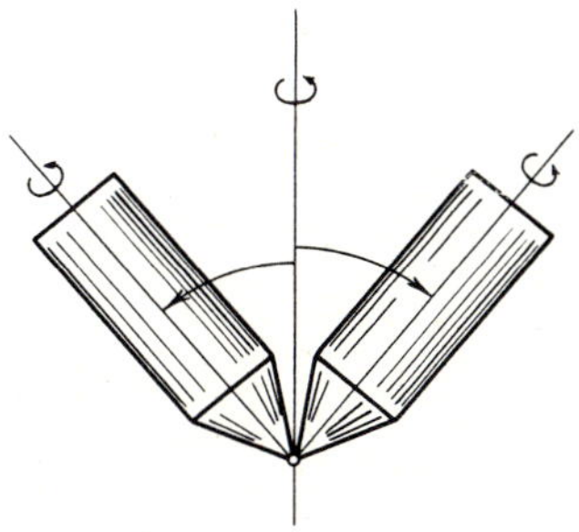

Fig.4

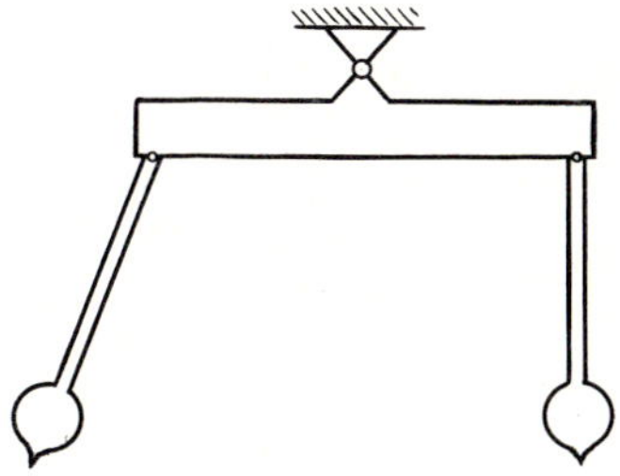

Fig.5

two pendulums, the first and the second (Fig.5). This system has three degrees of freedom and is nonlinear. Note that a rigid body with a fixed point also has three degrees of freedom. However, in our case there is the only first integral, vis. the integral of energy. Nevertheless they succeeded to exhaust this problem in some sence, the oscillations having had been not necessarily small. Professor V.F.Zhuravlyov applied the method of averaging when having been solving this problem, derived nonlinear equations and could integrate them. It is astonishing that the solution is expressed in simple trigonometric functions. You know the effect of Huygens: in the beginning one pendulum swings and the other is at rest, then it gradually begins to move and the first pendulum stops. Then the situation repeats. It looks like the periodical exchange of energy occurs.

The very derivation of equations of motion for multi-body systems causes significant difficulties. Our guest Professor J.Vittenburg [16] and Professor P.V.Kharlamov [17] with his pupils made a considerable break-through over these difficulties.

I would like to note that in mechanics and, perhaps, in many exact sciences general investigations compete with solving particular problemes. These particular solutions often appear to be rather important for understanding the essence of this or that branch of mechanics. Great Newton said not without reason that examples sometimes were more instructive than rules.

However the theory often is ahead of practical applications. To justify this thesis I mention the mechanics of non-holonomic systems, where there are a lot of theoretical investigations and constructions but rather few of practical examples. But nevertheless, theoretical and analytical research is very important. It takes a lot of time and demands a great analytical skill. It is well known that mathematicians made great efforts to investigate the problem of motion of a rigid body with a fixed point, but they obtained not so many practical results.
The new life was inspired to this problem in recent 40 years in connection with control of attitude of sattelites.

Surprising behaviour of mechanical systems,the so called paradoxes, play a great part in the development of mechanics. I would like to dwell on some of them. I am going to tell about one new chapter of mechanics of systems of bodies, about rotation of a rigid body on a string. In fact, it is a recently appeared branch of mechanics of a rigid body. The regularities of the motion of a rigid body with a fixed point and rolling of rigid bodies on a surface were thoroughly investigated earlier. But here all has begun from the experiment conducted in 1943 by the academician M.A.Lavrentyev and a remarkable experimeter from Kiev S.V.Malashenko [18]. They investigated the efficiency of the cumulative shell when it had been rotating. Usually the explosion destroys anything around, including an electric motor which puts the shell into rotation. S.V.Malashenko invented the means of hanging the shell on a string and protecting it (Fig.6*a*). And the following is just they had discovered.

While the angular velocity was comparatively small all was OK. Both the string and the axis of the model of the shell kept vertical directions. However, when the velocity became greater, the model unexpectedly deviated (Fig.6b). If one slightly increase the angular velocity and return the axis of the model to the vertical, it will perform stable rotation again. But if the angular velocity is getting greater the value will be met when the model and the string deviate on opposite sides (Fig.6*c*).

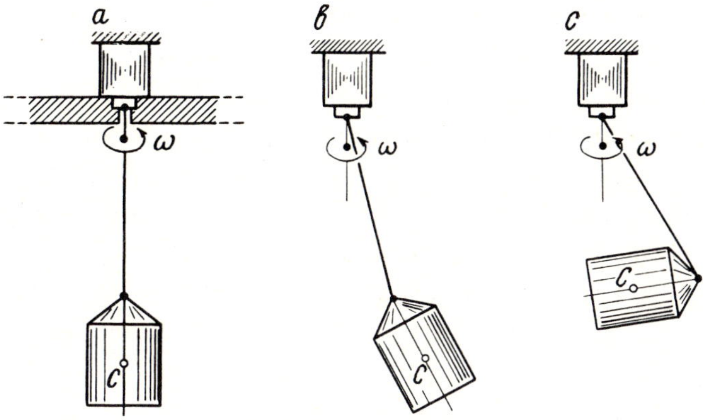

Fig.6

As it often occurs in mechanics, a new phenomenon is difficult to explain. But having been found the explanaton seems very simple. It appears that not a loss of stability but generation of the bifurcation takes place in this case. Moreover, and it is astonoshing, the new (first) mode of the steady-state motion is stable, but the old mode remains stable as well. The stability of the vertical in the sense of Lyapunov has been proved by E.P.Morozova and L.P.Smirnov [19], who are present at this session. M.E.Temchenko [20] from Kiev showed that the first bifurcation is also stable in the sense of Lyapunov and her colleague V.A.Storozhenko tried to find the Lyapunov function for the second mode. He had been searching for that function more than ten years but the full solution of this problem was found only three days ago. This mode does appear to be stable, excluded the small interval of the angular velocity. It is remarkable that the loss of stability for this mode was firstly established experimentally by S.V.Malashenko and then confirmed by another talented experimenter from Kiev P.G. Shishkin.

The motion of a rigid body on a string attracts a great attention.There are more than 200 papers devoted to this topic. New unexpected phenomena were noticed. Attempts were made to hang a body not at the point on its axis but close to it. When the angular velocity is sufficiently high the centre of mass of

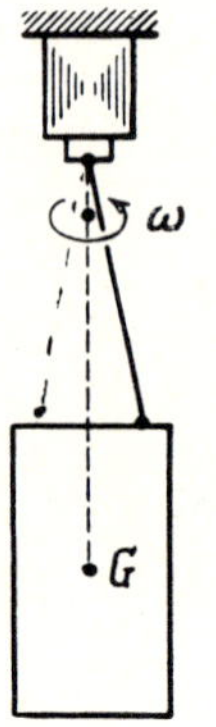

Fig.7

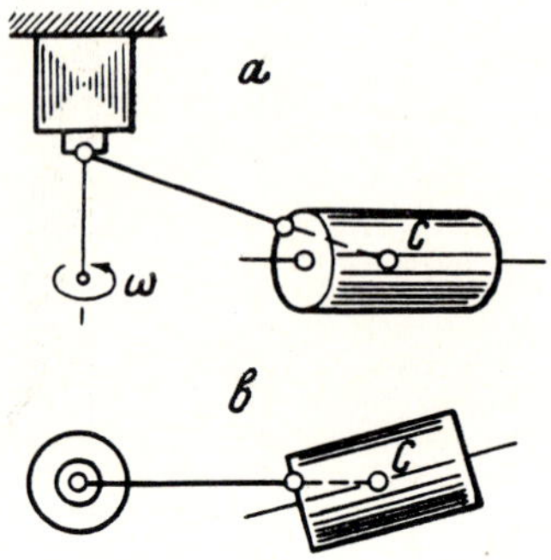

Fig.8

the body appears to lie on the vertical passing through the hanging point. The principal axis of inertia also finds itself vertically directed and the so called dynamic balancing takes place (Fig.7). But one has to start this body carefully, because the motion will meet several bifurcations. In particular, when the angular velocity increases, the angle of the axis of the body with respect to the horizon changes until at last this axis becomes horizontal. If the angular velocity increases further, the axis of the body remains horizontal unlike expected. But if you look from above you will see that the axis deviates from the plane containing the string and the centre of mass of the body (Fig.8). It is quite astonishing! This phenomenon can be easily explained but I have no time already to do it here.

The motion of a rigid body on a string can be also considered from another point of view. In fact, it is a limit case for the motion of two rigid bodies connected by a spherical hinge, one of the bodies being a string (Fig.9). This body gradually becomes the string as if loosing its mass. M.E.Temchenko showed that there is no essential difference between the motion of two bodies and the motion of a body on a string.

Here is one practical application of hanging a rigid body on a string. Fig.10 shows two centrifugal machines. One of them (the first, Fig.10 *a*) was constructed in Novosibirsk by B.V.

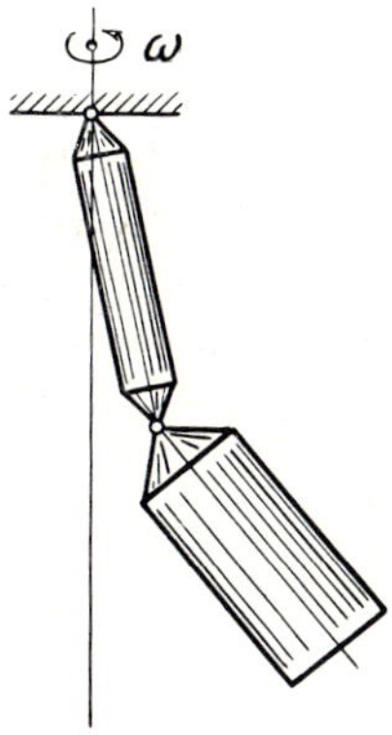

Fig.9

Voitsekhovsky and S.V.Malashenko. It is filled with a fluid and the rotation causes the artifical gravity which is ten or more times grater than the natural gravity. This device gives an example of the motion of a rigid body containing a fluid.

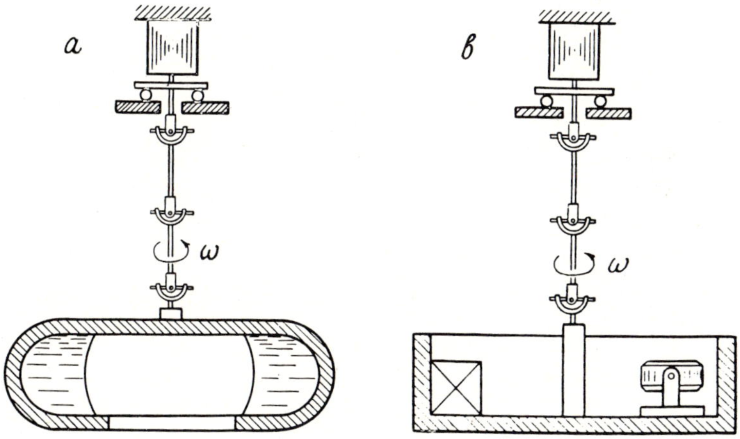

Fig.10

The motion of a rigid body with a fluid inside (Fig.11) is an interesting branch of mechanics. First investigations here treated a noncurled fluid. They were carried out by N.E.Zhukovsky and other scientists of the last century. In the forties of current century S.L.Sobolev made his remarkable research of the case when a top (a rigid body) and a fluid

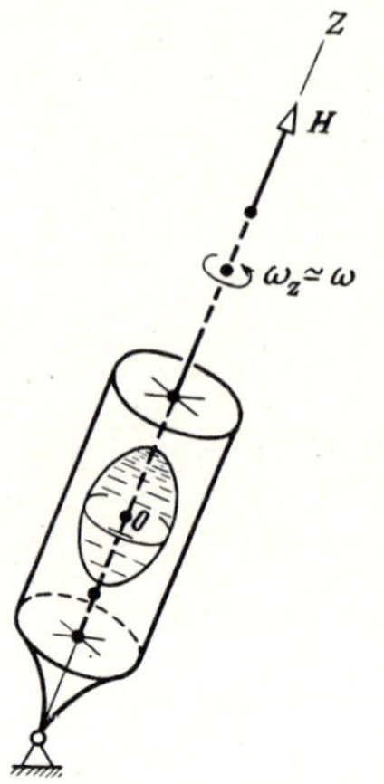

Fig.11

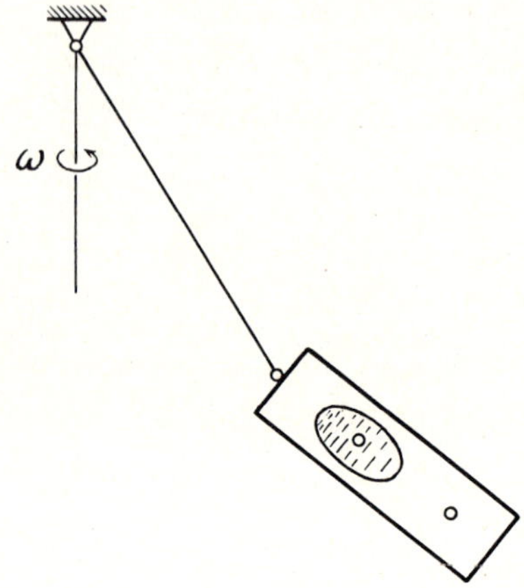

Fig.12

inside it rotate as a whole. S.V.Malashenko carried out experiments with such a top. The stability of the motion was also investigated. It is worth nothing that if the cavity has the shape of a cylinder, then the infinite spectrum arises, unlike the case of an ellipsoid when only two additional frequencies appear. It looks like a paradox. Of course, it is interesting to know what is the matter of fact in the case when a body with a fluid is hung on a string (Fig.12). It is a difficult problem but nevertheless it can be solved.

In Fig.10*b* you can see another centrifugal machine designed for testing gyroscopic and other devices for overload. This machine has five metres in diameter. It is hung on rods connected by Hook's hinges. In fact, it is a multi-body system. Now the centrifugal machine of ten metres in diameter has been designed.

Unfortunately I cannot dwell on many other problems, although very interesting. I would like only to mention that V.V. Rumyantsev, P.V.Kharlamov, V.A.Sarychev and their pupils dealt a lot with investigation of the motion of a rigid body on a string. The works by V.N.Skimmel [21] from Kazan and the scientists from Leningrad are also of great interest. As to dynamics of a rigid body with a fluid, this problem was treated by N.N.Moiseev and V.V.Rumyantsev [22] with their pupils, the researchers from Armenia, and pupils of the late academician S.L.Sobolev [23].

Now I would like to add some remarks on the role of computers in mechanics of rigid bodies. Certainly, computers in high extent help to solve problems arising in mechanics. Today they even can derive equatioins of motion and make analytical transformations of formulae. This is the so called computer algebra. In his introductory address academician K.V.Frolov told us about the Conference on nonlinear problems which was held last autumn at the Institute for Problems in Mechanics of the USSR Academy of Sciences.

I also delivered a lecture to that Conference and decided to intimidate the participans with the equations for one of the simplest problems of the motion of a rigid body on a string. Here they are.

$$ml^2(\ddot{\alpha} - \dot{\beta}^2\sin\alpha\cos\alpha) + mal\sin\alpha\,(\dot{\theta}^2\cos\theta + \ddot{\theta}\sin\theta) +$$
$$+ mal\cos\alpha\,\{[\ddot{\theta}\cos\theta - (\dot{\theta}^2 + \dot{\psi}^2)\sin\theta]\cos(\beta - \psi) +$$
$$+ (\ddot{\psi}\sin\theta + 2\dot{\theta}\,\dot{\psi}\cos\theta)\sin(\beta - \psi)\} + mgl\sin\alpha = 0,$$

$$ml\sin\alpha\,\{l(\ddot{\beta}\sin\alpha + 2\dot{\alpha}\,\dot{\beta}\cos\alpha) + a(\ddot{\psi}\sin\theta + 2\dot{\psi}\,\dot{\theta}\cos\theta)\,\times$$
$$\times\cos(\beta - \psi) + a[(\dot{\psi}^2 + \dot{\theta}^2)\sin\theta - \ddot{\theta}\cos\theta]\sin(\beta - \psi)\} = 0,$$

$$A(\ddot{\psi} + \ddot{\varphi}\cos\theta - \dot{\theta}\,\dot{\varphi}\sin\theta) + ma^2(\dot{\theta}\,\dot{\psi}\sin 2\theta + \ddot{\psi}\sin^2\theta) +$$
$$+ mal\sin\theta\,\{(\ddot{\beta}\sin\alpha + 2\dot{\alpha}\,\dot{\beta}\cos\alpha)\cos(\beta - \psi) +$$
$$+ [\ddot{\alpha}\cos\alpha - (\dot{\alpha}^2 + \dot{\beta}^2)\sin\alpha]\sin(\beta - \psi)\} - (A - C)\,\times$$
$$\times\{(\cos\theta\cos\delta - \sin\theta\cos\varphi\sin\delta)\,[\ddot{\psi}(\cos\theta\cos\delta -$$
$$- \sin\theta\cos\varphi\sin\delta) + \ddot{\theta}\sin\varphi\sin\delta + \ddot{\varphi}\cos\delta - 2\dot{\psi}\,\dot{\theta}\,\times$$
$$\times(\cos\theta\cos\varphi\sin\delta + \sin\theta\cos\delta) + 2\dot{\psi}\,\dot{\varphi}\sin\theta\sin\varphi\sin\delta] -$$
$$\dot{\theta}\,\dot{\varphi}\sin\theta(1 - 2\sin^2\varphi\sin^2\delta) - \dot{\theta}^2\sin\varphi\sin\delta(\cos\theta\cos\varphi\sin\delta +$$

$$+ \sin\theta\cos\delta) + \dot{\varphi}^2\sin\theta\sin\varphi\sin\delta\cos\delta\} = 0,$$

$$A(\ddot{\theta} + \dot{\psi}\,\dot{\varphi}\sin\theta) + ma^2(\ddot{\theta} - \dot{\psi}^2\sin\theta\cos\theta) +$$

$$+ mal\,\{[\sin\alpha\sin\theta + \cos\alpha\cos\theta\cos(\beta - \psi)]\,\ddot{\alpha} -$$

$$- \ddot{\beta}\sin\alpha\cos\theta\sin(\beta - \psi) + \dot{\alpha}^2\cos\alpha\sin\theta - (\dot{\alpha}^2 + \dot{\beta}^2) \times$$

$$\times \sin\alpha\cos\theta\cos(\beta - \psi) - 2\dot{\alpha}\,\dot{\beta}\cos\alpha\cos\theta\sin(\beta - \psi)\} +$$

$$+ (A - C)\,\{\sin\varphi\sin\delta\,[\ddot{\psi}(\sin\theta\cos\varphi\sin\delta - \cos\theta\cos\delta) -$$

$$- \ddot{\theta}\sin\varphi\sin\delta - \ddot{\varphi}\cos\delta] - \dot{\psi}\,\dot{\varphi}\,[\cos\theta\cos\varphi\sin2\delta +$$

$$+ \sin\theta\,(1 - 2\cos^2\varphi\sin^2\delta)] - \dot{\theta}\,\dot{\varphi}\sin2\varphi\sin^2\delta -$$

$$- \dot{\varphi}^2\cos\varphi\sin\delta\cos\delta + \dot{\psi}^2(\sin\theta\cos\varphi\sin\delta - \cos\theta\cos\delta) \times$$

$$\times (\cos\theta\cos\varphi\sin\delta + \sin\theta\cos\delta)\} + mga\sin\theta = 0,$$

$$A(\ddot{\varphi} + \ddot{\psi}\cos\theta - \dot{\psi}\,\dot{\theta}\sin\theta) + (A - C)\,\{[\ddot{\psi}(\sin\theta\cos\varphi\sin\delta -$$

$$- \cos\theta\cos\delta) - \ddot{\theta}\sin\varphi\sin\delta - \ddot{\varphi}\cos\delta]\cos\delta + \dot{\theta}\,\dot{\psi} \times$$

$$\times [\sin\theta\,(1 - 2\sin^2\delta\cos^2\varphi) + \cos\theta\cos\varphi\sin2\delta] +$$

$$+ \dot{\theta}^2\sin\varphi\cos\varphi\sin^2\delta + \dot{\psi}^2\sin\varphi\sin\theta\sin\delta \times$$

$$\times (\cos\theta\cos\delta - \sin\theta\cos\varphi\sin\delta)\} = 0\,.$$

D.M.Klimov and his pupil V.M.Rudenko checked those equations on a computer and caught two mistakes. Fortunately, they turned out to be not mistakes but misprints. Thus the skill of calculation is still high in the USSR. Of course, one cannot demand of a computer that it cannot do. It cannot choose the method of research, find bifurcations, invent the adequate change of variables and do many other things which the human brain performs.

I am not doubting that mechanics of multi-body systems is the branch which will long develop. Apart from purely steady-state motions and their stability, which are almost exclusively treated now, the transient processes and non-steady motions will be the subject of future investigations. The development of new branches of sciences often inspire a new life to old traditional problems which seem to be fully exhausted.

That was a concise survey of the development of investigations in dynamics of systems of rigid bodies. I have told you some basic results as well as problems which are still to be solved. I thought to tell you two times more. I hope to return to this theme further. Thank you very much for the attention.

References

1. Николаи Е.Л. О движении уравновешенного гироскопа в кардановом подвесе // ПММ. 1939. Т.3, вып.4. С.3-34.

2. Четаев Н.Г. Устойчивость движения. Работы по аналитической механике. М.: Изд-во АН СССР, 1962. 535 с.

3. Magnus K. Beiträge zur Dynamik des Kräftefreien, Kardanisch gelagerten Kreisels // ZAMM. 1955. Bd.35. H.1/2. S. 23-34.

4. Plymale B.T. and Goodstein R. Nutation of a free Gyro subjected to an impulse. J. Appl. Mech., 1955. Vol. 22, 3. C.365-366.

5. Климов Д.М., Степаненко Н.П. Об интегрировании уравнений движения гироскопа в кардановом подвесе // Инж.ж. МТТ. 1967. 6. С.143-150.

6. Лунц Я.Л. О неустойчивости оси фигуры гироскопа // ПММ. 1960. Т.24, вып.4. С.763-765.

7. Брюно А.Д. О движении гироскопа в кардановом подвесе //Изв. АН СССР. МТТ. 1972. 6. С.5-18.

8. Валеев К.Г. О прецессии симметричного гироскопа в кардановом подвесе// Механика твердого тела. Киев: Наук. думка, 1974. Вып. 7. С.28-36.

9. Журавлев В.Ф. К вопросу об оценках эффекта Магнуса // Докл. АН СССР. 1976. Т.226, 3. С.541-543.

10. Ишлинский А.Ю. Ориентация, гироскопы и инерциальная навигация. М.: Наука, 1976. 670 с.

11. Харламов П.В. Некоторые классы точных решений задачи о движении системы гироскопов Лагранжа // Механика твердого тела. Киев: Наук. думка, 1987. Вып.19. С.42-54.

12. Schiehlen W. Nonlinear phenomena in multibody system dynamics// Delft Progress Report. 1985. 10. S.105-120.

13. Степаненко Н.П. Об уходе двух связанных гироскопов // Изв. АН СССР. МТТ. 1969. 1. С.40-44.

14. Рогачева Л.Н., Степаненко Н.П. О движении двух твердых тел, соединенных сферическим шарниром // Вестн. МГУ. Сер. 1. Математика, механика. 1971. 6. С.106-111.

15. Лесина М.Е. О колебаниях оси маховика в теле-носителе // Механика твердого тела. Киев: Наук. думка, 1979. Вып. 11. С.32-37.

16. Виттенбург Й. Динамика систем твердых тел / Под ред. В.В. Румянцева. М.: Мир, 1980. 292 С.

17. Харламов П.В. Об уравнениях движения системы твердых тел // Механика твердого тела. Киев : Наук. думка, 1972. Вып. 4. С. 52-73.

18. Малашенко С.В. Некоторые экспериментальные исследования, относящиеся к вращению тел // ПМТФ. 1960. 3. С. 136-143.

19. Морозова Е.П. Об устойчивости вращения твердого тела, подвешенного на струне// ПММ. 1956. Т. 20, вып. 5. С. 621-626.

20. Темченко М.Е. Об устойчивости одного из положений динамического равновесия одной механической системы// Докл. АН СССР. 1957. Т. 117, 1. С. 50-52.

21. Скимель В.Н. О движении гиростата, подвешенного на струне. Тр. Межвуз. конф. по прикл. теории устойчивости движения и аналит. механике, 1962. Казань: Казан. авиац. ин-т, 1964. С. 118-122.

22. Моисеев Н.Н. и Румянцев В.В. Динамика тела с полостями, содержащими жидкость. М.: Наука, 1965. 439 с.

23. Соболев С.Л. О движении симметричного волчка с полостью, наполненной жидкостью // ПМТФ. 1960. 3. С. 20-55.

24. Климов Д.М., Харламов С.А. Динамика гироскопа в кардановом подвесе. М.: Наука, 1978. 208 с.

25. Белецкий В.В. Двуногая ходьба. Модельные задачи динамики и управления. М.: Наука, 1984. 286 с.

$$\frac{d^2\zeta}{d\tau^2} - \frac{3}{2}\sigma \frac{\left(1 + e\cos\nu\right)^3}{\left(1 - e^2\right)^3} \sin 2\left(\beta\tau - \nu + \zeta\right) +$$

$$+ \chi\kappa \frac{\left(1 + e\cos\nu\right)^6}{\left(1 - e^2\right)^6}\left(\beta - \frac{d\nu}{d\tau} + \frac{d\zeta}{d\tau}\right) = 0. \tag{11}$$

To obtain the solution of this equation the averaging method was used. If 2β is not integer then

$$\frac{d^2\zeta}{d\tau^2} + \chi\kappa I_1(e)\frac{d\zeta}{d\tau} = -\chi\kappa\left(\beta I_1(e) - I_2(e)\right),$$

$$I_1(e) = \frac{1 + 3e^2 + \frac{3}{8}e^4}{\left(1 - e^2\right)^{9/2}}, \tag{12}$$

$$I_2(e) = \frac{1 + \frac{15}{2}e^2 + \frac{45}{8}e^4 + \frac{5}{16}e^6}{\left(1 - e^2\right)^6}.$$

We can integrate this equation easily. All ring motions approach to the motion with the angular velocity

$$\omega^* = \frac{I_2(e)}{I_1(e)}\omega_0 .$$

If $2\beta = m$ (integer), then instead of (12) we have the equation

$$\frac{d^2\zeta}{d\tau^2} - \frac{3}{2}\sigma\,\Phi_m(e)\sin 2\zeta + \chi\kappa I_1(e)\frac{d\zeta}{d\tau} =$$

$$= -\chi\kappa\left(\frac{m}{2}I_1(e) - I_2(e)\right), \tag{13}$$

$$\Phi_m(e) = \frac{1}{2\pi}\int_0^{2\pi}\frac{\left(1 + e\cos\nu\right)^3}{\left(1 - e^2\right)^3}\cos\left(m\tau - 2\nu\right)d\tau .$$

If $\chi\,\kappa\,\left|\,m\,I_1 - 2\,I_2\right| > 3\,\sigma\,\left|\,\Phi_m\right|$, then any $\omega \longrightarrow \omega^*$. If $\chi\,\kappa\,\left|\,m\,I_1 - 2\,I_2\right| < 3\,\sigma\,\left|\,\Phi_m\right|$, then either $\omega \longrightarrow m\,\omega_0/2$ or $d\zeta/d\tau$ is the Π-periodical function of ζ.

Dynamics of symmetrical viscoelastic body.

The equations of motion for the body were derived for the case of longitudinal oscillations [3, 8]. There are three types of precession: cylindrical, hyperbolical and conical.

The cylindrical precession can be either stable or unstable.

The hyperbolical precession is asymptotically stable if $A_3 > A_1$. The conical precession is unstable if $A_3 > A_1$ and asymptotically stable if $A_3 < A_1$,

$$\sin^2\theta_* \neq \frac{4\,A_1^2 - 6\,A_1A_3 + 3\,A_3^2}{3\left(2\,A_1 - A_3\right)\left(4\,A_1 - 5\,A_3\right)} .$$

θ_* is the angle between the axis of symmetry of the body and vector $\bar{\gamma}$ in the case of conical precession.

If the characteristic time of the body motion about mass center much less than the orbital time then one can introduce a small parameter $\mu = A_1\,\omega_0\,/\,I_2^0$. Using averaging method body motion equations were derived [3].

$$\frac{d\delta_1}{d\tau} = \frac{1}{64}\sin\delta_1\cos\delta_1\sin^2\delta_2\left[\sin^2\delta_2\left(1 + 3\cos^2\delta_1\right) + \right.$$
$$\left. + 12\sin^2\delta_1\cos^2\delta_2\right], \qquad (14)$$

$$\frac{d\delta_2}{d\tau} = -\frac{1}{64}\sin\delta_2\cos\delta_2\left[\sin^2\delta_2\left(3\cos^4\delta_1 + \right.\right.$$
$$\left.\left. + 2\cos^2\delta_1 + 3\right) + 4\sin^2\delta_1\cos^2\delta_2\left(1 + 3\cos^2\delta_1\right)\right],$$

$$\tau = 72\,\chi\,b\,\omega_0^4\,\frac{\left(2\,A_1 - A_3\right)}{A_1^2}\sum_{n=1}^{\infty}\left(\frac{H_{33}^{(n)}}{\omega_n}\right)^2 ,$$

$$\cos\delta_1 = \frac{I_3}{I_2}, \qquad \cos\delta_2 = \frac{I_1}{I_2} .$$

I_1, I_2, I_3 are the Andoyer's canonical variables.

Let $I = 0$, then $\delta_2 = \pi/2$ and

$$\frac{d\delta_1}{d\tau} = \frac{1}{64} \sin \delta_1 \cos \delta_1 \left(1 + 3 \cos^2\delta_1\right) .$$

The last equation has two unstable equilibrium points $\delta_1 = 0, \pi$ and the asymptotically stable equilibrium point $\delta_1 = \pi/2$.

If $I_1 \neq 0$ then the system (14) has the family of equilibrium points: $\delta_2 = 0, \pi$, $\delta_1 = \delta_1^*$ (arbitrary constant). All the points are asymptotically stable relative to the variable δ_2 and stable relative to the variable δ_1.

Dynamical problems of viscoelastic shells.

The evolution of fast rotations of cylindrical shell were investigated in [9]. It was shown that the angular momentum of the shell tends to move towards the orbital plane.

Depending on the shell parameters the two final motions take place: (i) the symmetry axes of the shell lies in the orbital plane; (ii) the symmetry axes of the shell is perpendicular to the orbital plane. The instantaneous rotation axes lie in orbital plane in both cases.

The investigation of the motion of a spherical shell is given in [10].

Certain problems of shallow spherical shells are solved in [11]. The shell has stationary rotation with angular velocity r_3 when symmetry axes of the shell is perpendicular to orbital plane. The rotation is stable if $r_3 > \omega_0/2$ or $r_3 < - \omega_0$ and unstable if $- \omega_0 < r_3 < \omega_0/2$ for a case of a rigid body [4, 5]. This rotation is asymptotically stable if $r_3 > \omega_0/2$ and unstable if $r_3 < \omega_0/2$ for the case of a viscoelastic body [11].

References

1. Santiny, P.: Stability of flexible spacecrafts. Acta Astronautica, 1976, vol. 3, N 9-10, p. 685-713.
2. Маркеев, А.П.: К динамике упругого тела в гравитационном поле. Космические исследования, 1989, Т. 27, вып. 2, с. 163-175.

3. Маркеев, А.П.: Влияние продольных упругих колебаний тела на его быстрые вращения в гравитационном поле. Известия АН СССР. Механика твердого тела, 1989, N 6, с. 38-45.

4. Черноусько, Ф.Л.: Об устойчивости регулярной прецессии спутника. Прикладная математика и механика, 1964, Т. 28, N 1, с. 156-157.

5. Маркеев, А.П.: Резонансные эффекты и устойчивость стационарных вращений спутника. Космические исследования, 1967, Т. 5, N 3, с. 365-375.

6. Климов, Д.М.; Маркеев, А.П.; Холостова, О.В.: К динамике упруговязкого кольца в гравитационном поле. Препринт Ин-та проблем механики АН СССР, N 406, М.: 1989.

7. Климов, Д.М.; Маркеев, А.П.: Динамика неоднородного упругого кольца в гравитационном поле. Препринт Ин-та проблем механики АН СССР, N 331, М.: 1988.

8. Маркеев, А.П.: Об одном частном случае движения динамически симметричного упруговязкого тела в центральном ньютоновском гравитационном поле. Космические исследования, Т. 28, 1990.

9. Маркеев, А.П.: Эволюция быстрых вращений вязкоупругой цилиндрической оболочки в гравитационном поле. Известия АН СССР. Механика твердого тела, 1990.

10. Холостова, О.В.: О быстрых вращениях упругой сферической оболочки в гравитационном поле. Известия АН СССР. Механика твердого тела, 1990.

11. Карпов, И.И.; Климов, Д.М.; Маркеев, А.П.: Аналитический вывод на ЭВМ уравнений движения упругого тела в гравитационном поле. Препринт Ин-та проблем механики АН СССР, N 411, М.: 1989.

The Dynamics of Broadband Active Vibration Control Systems with Controllable Damping for Machines and Mechanical Structures

S.V. KRAVCHENKO, K.V. FROLOV

Mechanical Engineering Research Institute,
USSR Academy of Sciences,
4, Griboedova, 101830, Moscow, USSR

Summary

The theory of active vibration control based on the principle of broadband compensation coupled with controllable vibration damping is presented. The applications of this principle to linear, parametric and nonlinear mechanical structures are discussed. For the first time the possibility of the active vibration control with the aid of the stochastic self-excited vibration spectrum transformation nonlinear phenomena is investigated. The theoretical and experimental results are compared for various complex mechanical objects.

Introduction

The problem of minimizing the link dynamic force reactions with the aid of controllable active force generators placed between the link connection points is solved for different mechanical systems. Assume that the passive mechanical link between the two oscillatory systems has the arbitrary structure; it is described by the transmissibility function $\Phi(t)$. If the active generator creates the force $f(t)$, than the link dynamic force is expressed as

$$R(t)=\Phi(t)[y_1(t)-y_2(t)]-f(t), \tag{1}$$

where $y_1(t)$ and $y_2(t)$ are the link connection points displacements. According to Eq.1, it is necessary for minimizing the general link dynamic force reaction $R(t)$ to create the active force $f(t)=\Phi(t)[y_1(t)-y_2(t)]$. This conclusion corresponds to the invariance principle of the control theory [1], which states that for all types of transmissibility functions, the link force reaction $R(t)$ can be fully compensated with the

N.V. Banichuk · D. M. Klimov · W. Schiehlen (Eds.)
Dynamical Problems of Rigid-Elastic
Systems and Structures
IUTAM Symposium Moscow, USSR, May 23-27, 1990

aid of additional control loop described by $-\Phi(t)$. However, the physical realization of this broadband vibration compensation principle is limited by the closed-loop control system instability and active force generators technical parameters. Moreover, it is impossible to reduce the resonant vibration amplitudes with the aid of the broadband vibration compensation principle. So, it is suggested, for increasing the control system stability and reducing resonant vibration amplitudes, to create additionally the narrow band controllable damping forces using the special narrow band filters in the additional control loop. The coupling of broadband compensation and controllable damping principles in the active vibration control systems for the first time was suggested by Frolov [2].

<u>Linear systems</u>

Let us consider the elastic vibrating beam with two active vibration isolators at the beam ends (Fig.1). The active forces f_1 and f_2 needed for vibration isolators reactions broadband compensation and controllable damping of beam natural resonant oscillations are formed by the signals from the force transducers placed inside the passive vibration isolators. The transfer functions W_t, W_c, W_d, W_e describe transducers, compensation channels, controllable damping channels and exciters, respectively. The loops of controllable damping W_d include the parallel narrow-band filters with structure and optimal parameters chosen according to the fixed points method [3]. Theoretical investigation taking into consideration ten first beam normal modes showed that high vibration isolation effectiveness and self-excitement elimination at all investigated frequencies can be achieved with the aid of the narrow band controllable damping.

Fig.2 shows the experimental force transmissibilities for the beam isolators under the broad band random excitation. It is seen from Fig.2 that the force transmissibility reduction by 10 to 12 dB in the wide frequency range, and more then 20 dB suppression of the first beam natural frequency resonant peaks are obtained due to the combination of compensation and damping control principles.

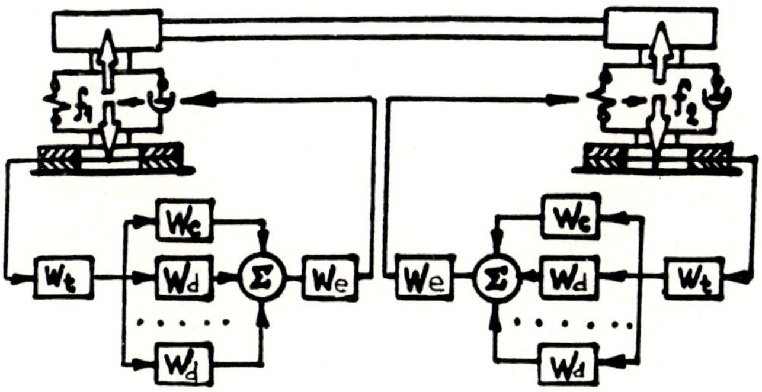

Fig.1. Active vibration control of the linear elastic beam

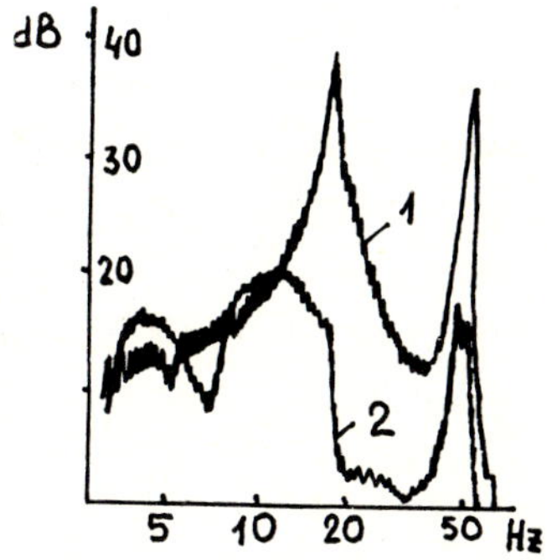

Fig.2. Experimental transmissibilities: 1-passive isolation, 2-active vibration compensation with controllable damping

Parametric systems

The history of the investigations concerned with machines and structures parametric vibration problems is a fairly old one [4]. Examples of such problems are vibrations of gear mechanisms, cyclic machines, rotor systems with bearing supports, elevators, pipelines and tubes conveying fluid, etc. For all these and many others parametrically excited technical objects, active vibration control methods and means could be used. The most simple model of all these examples is the single-degree-of-freedom vibratory system with parametric stiffness $(k+\Delta k\cos\Omega t)$. For this system (Fig.3) we shall investigate the active vibration control based upon the coupling of compensation and damping control principles. Assuming that the sensor of the control system is the force transducer placed under the passive vibration isolator (Fig.3), the nondimensional form the equation of motion under the unit force excitement is given by

$$[p^2+(\lambda p+(1+a\cos\Omega\tau))(1-W(p))]x=\cos\omega t \qquad (2)$$

where x is the object displacement, $\lambda=h/m\omega_0$ is the damping ratio, $\omega_0 = \sqrt{k/m}$ is the natural frequency, f_0 and ω_0 are exciting force amplitude and frequency, respectively.

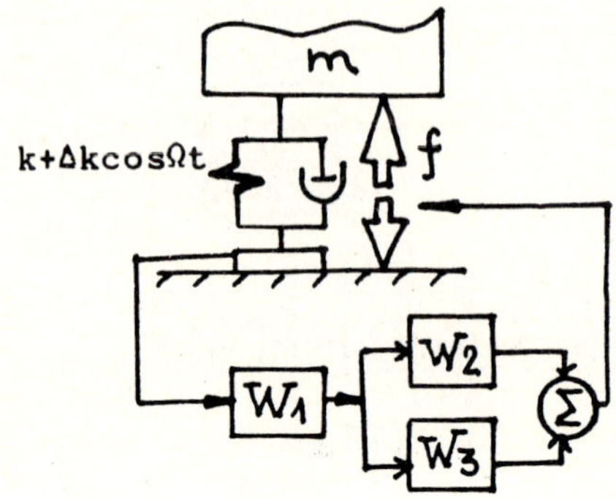

Fig.3. Schematic of the active vibration control in the parametric oscillatory system

The control system transfer function is written as [2]:

$$W(p)=W_1(p)[W_2(p)+W_3(p)],$$

$$W_1(p)=p/(p+r_1);\ W_2(p)=k_2;\ W_3(p)=k_3r_3/(p_3^2+pr_3/Q_3+r_3^2), \qquad (3)$$

where W_1 describes the force transducer including the low frequency filter with cutoff frequency r_1; W_2 is the transfer function of the broadband compensation channel with the gain k_2; W_3 describes the simplest controllable damping channel presented by the 2nd order low-pass filter with the gain k_3,quality factor Q_3 and cutoff frequency r_3.

For modulation parameter a=1, the effect of the active vibration control on the stability boundaries was investigated using Floquet method. In Fig.4 the relationships between the non-dimensional frequency Ω/ω_0 and the compensation gain k_2 are compared for different amounts of controllable damping. Fig.4a shows the stable regions without controllable damping (k_3=0). It can be seen that the upper stability bound is at k_2^*=0,7. It is evident from Eq.1 that the effectiveness of active vibration compensation control increases when k_2 tends to 1. The widest domain of stability corresponds to the optimal narrow band controllable damping (Fig.4b) [3]. In this case the upper stability bound reached its maximum value at k_2^*=0,9 and the instability regions practically disappeared. Hence, optimal controllable damping increases the effectiveness of the

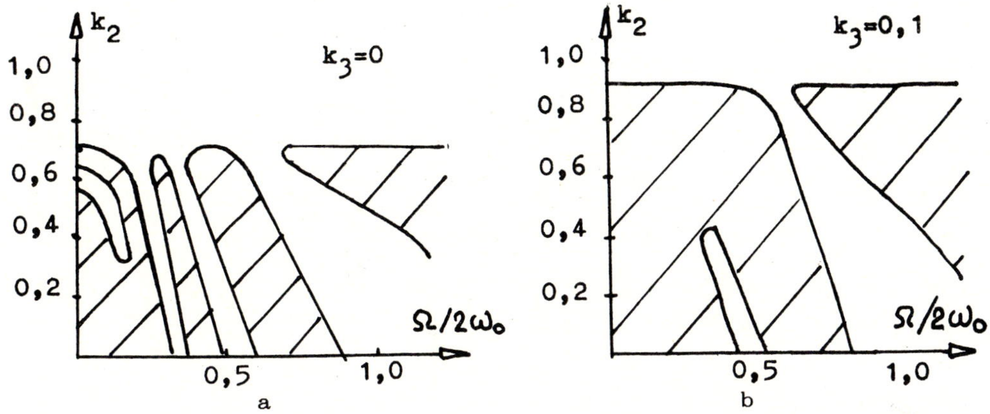

Fig.4. The stability domains (hatched) for the active control of the parametric oscillatory system

broadband force compensation and at the same time reduces the potentials for parametric instability.

Fig.5 shows the isolator force amplitude frequency response for the system with parametric stiffness. The amplitude peaks of curve 1 are due to parametric resonances. The effectiveness of broadband force compensation and controllable damping in the wide frequency range is about 10 dB. Moreover, the parametric resonance frequencies are shifted according to the formula $\Omega/2\omega_0=\sqrt{(1-k_2)/n}$ (where n=1,2,3,...), and the vibration amplitudes reduction at these frequencies is sufficiently high.

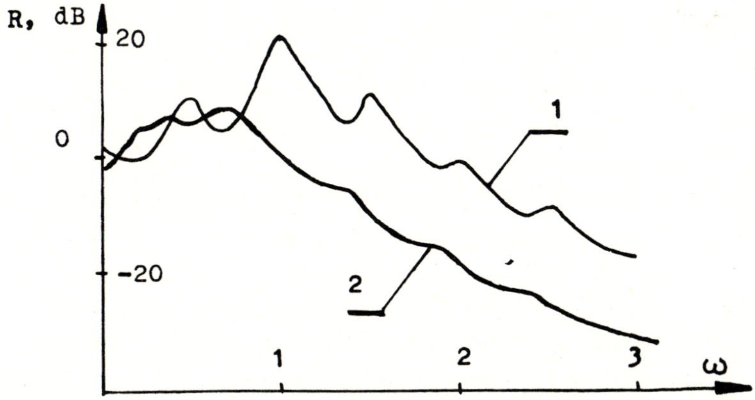

Fig.5. Force transmissibility of the system with parametric stiffness. 1 - passive vibration isolation, 2 - active control

Nonlinear systems (periodic vibration)

Study of nonlinear phenomena in the active vibration control is of great importance as it has many practical applications [5]. Nonlinear features of the vibration objects and control systems can play both positive and negative role in the active vibration control using controllable damping and compensation principles.

Let us consider the periodic vibration at the simple one dimensional active vibration control system with the cubic passive isolator stiffness coefficient $c(x)=\gamma x^3$. This model describes the supports often used in the vibration isolation systems. In this case the control system is described by equations (3), but the parametric stiffness coefficient is replaced by the nonlinear expression c(x) in the equation of motion (2). The controllable damping and broadband vibration compensation effect on the main nonlinear periodic vibration was investigated by the harmonic linearization method. Assuming $W_1=1$ and neglecting the passive isolator dissipation, the optimal parameters of the controllable damping could be found via the fixed points method [3]. The minimization of the main resonance amplitudes was achieved using the optimal parameters:

$$r_{3opt}=\sqrt{(1+q)(1-k_2)}, \qquad Q_{3opt}=\sqrt{2(1-k_2)/3k_3} \tag{4}$$

where $q=0.75\gamma A^2$ is the harmonic linearization coefficient, A is the displacement amplitude of the nonlinear vibration. Fig.6 shows the comparison of the vibration effectiveness for the different sets of control parameters. Under the optimal controllable damping, the maximum vibration amplitude in the nonlinear active vibration control system is given by

$$a_f=\sqrt{[1+\sqrt{1+3\gamma/\sqrt{(1-k_2)k_3/2}}\,]/1,5\gamma} \tag{5}$$

The amplitude a_f corresponds to the fixed points ω_f [3], i.e. to the frequencies independent of the nonlinear oscillatory system damping. Eq.5 yields the safe distance between the supports which guarantees the absence of impacts during vibration.

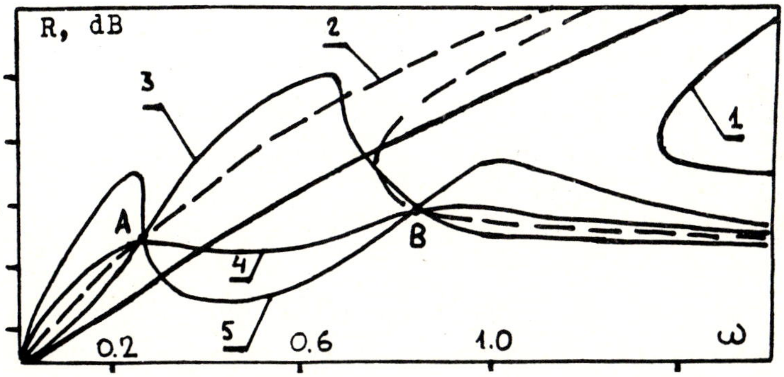

Fig.6. Force transmissibilities of the non-linear system. 1 - passive vibration isolation, 2 - active broadband vibration compensation (k_2 =0.7, k_3=0); 3,4,5 - active compensation coupled with controllable damping (k_2=0.7, k_3=0.15, r_3=1.0):3 - Q_3=2; 4 - Q_3=1 (optimal value);5 - Q_3=0.5. A,B- fixed points

Active vibration control effectiveness for a nonlinear system may be reduced by the combined vibration excitation. The subharmonic and superharmonic oscillations in the active isolation system using the aforementioned principles were investigated by the harmonic balance method. For instance, Fig.7 shows the magnitude of the force isolators transmissibility vs frequency for the main harmonic vibration (R_1) and the third order superharmonic vibration (R_3). The calculation has been performed for γ=0.1 under the conditions of stable motion

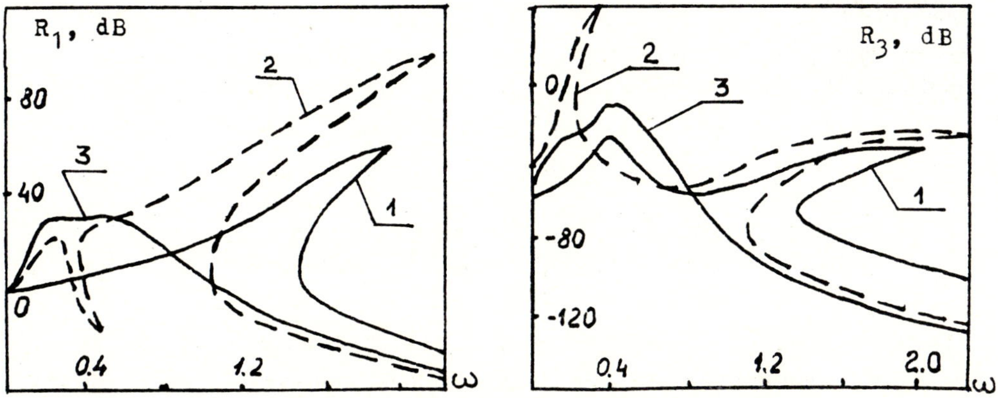

Fig.7. Main harmonic and superharmonic oscillations:1 - passive vibration isolation; 2 - active force compensation; 3 - coupling of optimal controllable damping and force compensation.

It is seen from Fig.7, that the application of active vibration compensation alone shifts the peak of frequency response to the low frequency range, and the main harmonic (R_1) and superharmonic (R_3) peaks increase considerably. The optimal controllable damping (4) practically eliminates combined vibration excitation and minimizes all harmonic amplitudes.

Nonlinear systems (stochastic self-excited vibration)

Sometimes the nonlinearities in the feedback channels can increase the vibration isolation effectiveness by transformation of the discrete vibration spectrum to the wide-band one in the state of stochastic self- excitement [6].

Assuming the linear passive vibration isolator characteristics and introducing the additional nonlinear amplifier W_4, the nondimensional equations of motion are given by

$$\ddot{x}+\lambda\dot{x}+x=W_4(z)+\sum f_i^0\cos\omega_i t$$

$$\dot{y}+r_1 y=\lambda\dot{x}+x \tag{6}$$

$$\ddot{z}+(r_3/Q_3)\dot{z}+r_3^2 z=k_2\ddot{y}+k_2(r_3/Q_3)\dot{y}+r_3^2(k_2+k_3)y$$

where $\lambda, r_1, r_3, Q_3, k_2, k_3$ are the parameters of the mechanical object and control system, f_i^0 and ω_i are amplitude and frequency of the harmonic excitation. The nonlinear amplifier characteristic $W_4(z)$ is given by

$$W_4(z)=\begin{cases} z, & \text{if } |z|\le\zeta \\ \zeta\,\mathrm{sign}z, & \text{if } |z|>\zeta \end{cases} \tag{7}$$

where ζ is the threshold parameter of the nonlinear amplifier. Numerical investigation of the Eqs.(6),(7) yields the wide ranges of parameters where the stochastic self-excited oscillations exist instead of periodic one [7]. The energy from the narrow frequency bands (discrete spectrum) is redistributed to the wide frequency range (continuous spectrum). This effect illustrates the transfer from periodic to irregular behavior; it was investigated via the nonlinear vibration isolation force transmissibility R(t) realization analysis. Eqs.(6),(7) were integrated using fourth order Runge-Kutta method. The local maximum points mapping described by the equation $R^{n+1}_{max}=\psi(R^{n}_{max})$ was considered. This mapping is the analog of the well known Poincare map [7]. It was used by E. Lorenz in his pioneer work on the stochastic behavior of the simple systems [8]. The stochastic motion was not found for the active vibration control systems with the broadband compensation of vibration forces alone. In this case the nonlinear system is nearly conservative one, and only multi-periodic vibration with the discrete spectrum was found in the wide range of parameters (Fig.8a) .The nonlinear active vibration control system began

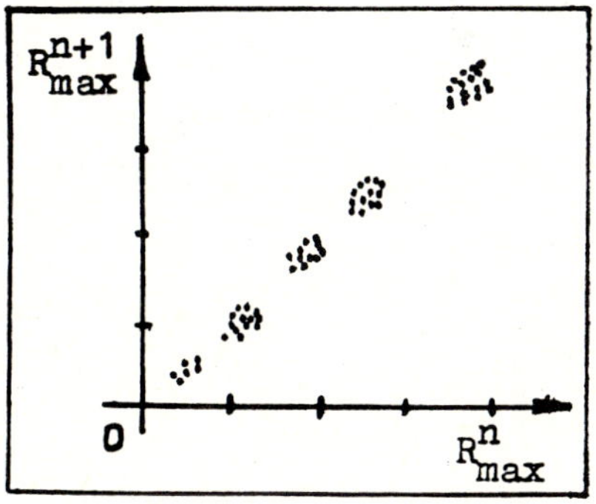

a

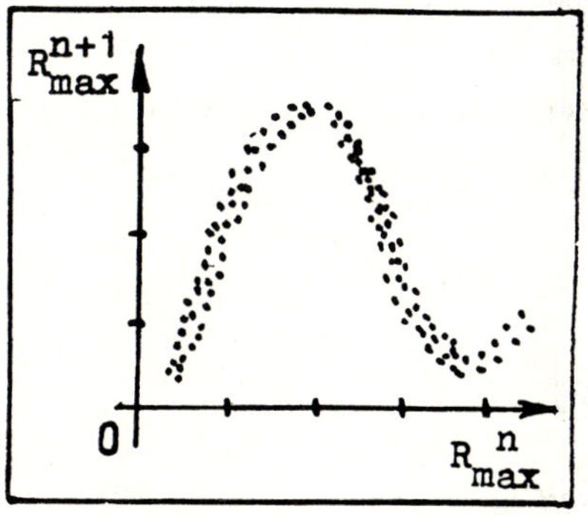

b

Fig.8. Local maximum points mappings. a - multi-periodic vibration, b - stochastic self-excited vibration

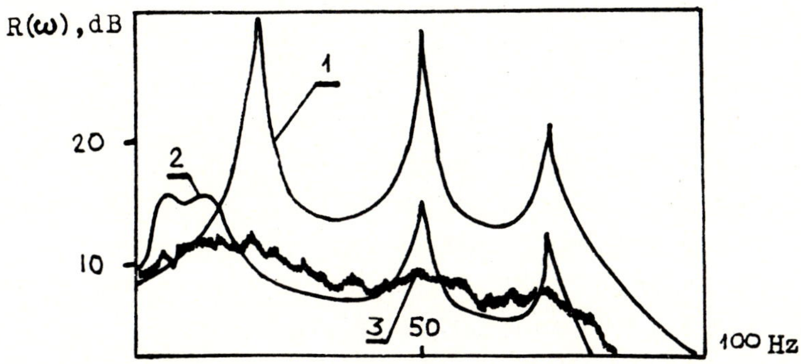

Fig.9. Experimental spectra of force transmissibilities. 1 - passive isolation, 2 - linear active vibration control, 3 - non-linear active control

to behave as the stochastic one when the optimal controllable damping (4) was used (i.e. all main and combined resonances were suppressed). The local maximum map ψ (Fig.8b) could be approximated by a one-dimensional transformation. Recent investigations have revealed that stochastic behavior occurs if a dynamic system can be reduced to the one-dimensional mapping of the interval into itself [7].

Experimental results (Fig.9) show the use of the stochastic self-excited vibration spectrum transformation nonlinear phenomena in the vibration control.

Conclusions

Universal active vibration control methods are proposed for linear, parametric and nonlinear mechanical systems. It is shown that the increase of the vibration isolation effectiveness, suppression of resonance peaks and reducing of instable regions can be achieved by coupling of vibration broadband compensation and controllable damping. Hence it is possible to use the proposed control principles not only for vibration isolation but also for dynamic processes stabilization.

References

1. Pervozvansky, A.A. Automatic control. Moscow: Nauka 1986 (in Russian).

2. Frolov, K.V.: The increase of the resonance peaks in the oscillatory system by the mechanical parameters control. Sov.Mach.Sci.(1965) no.3, 38-42.(Translated from Russian.)

3. Kravchenko, S.V.: The active vibration control synthesis by using the fixed points method. Sov.Mach.Sci. (1986) no.3 22-27.(Translated from Russian.)

4. Frolov, K.V.: Parametric problems of machine dynamics. Rev.Roum.Sci.Techn.-Mech.Appl. (1972) 17 no.2, 265-290.

5. Kolovsky, M.Z. Automatic vibration control., Moscow: Nauka, 1976. (in Russian).

6. Kravchenko, S.V.et al.: The active vibration control method. USSR Patent no.1348852, 1987 (in Russian).

7. Schuster, H.G. Deterministic Chaos. An Introduction. Physic Verlag Gmbh. Weinheim, 1984.

8. Lorenz, E.N.: Deterministic Nonperiodic Flow. J.Atmosph.Sci 20 (1963) 130-141.

Response Localization in Disordered Periodic Structures

Y. K. Lin and G. Q. Cai

Center for Applied Stochastics Research, Florida Atlantic University
Boca Raton, Florida 33431, USA

Summary

Due to material and manufacturing variabilities, a structure designed to be spatially periodic cannot be exactly periodic. The departure from exact periodicity is known as disorder. Under excitation, the response of a disordered periodic structure is much more confined as compared with its ideally periodic counterpart. Assuming that disorder can be described in terms of probability or statistics, the average exponential decay rate of the response amplitude with respect to the distance is investigated using a perturbation procedure.

Introduction

An ideal periodic structure is composed of identically constructed cell-units to form a spatially periodic array. It is well-known that such a structure has the common property of alternate wave-passage and wave-stoppage frequency bands [e.g. 1,2]. If an ideal periodic structure is undamped and is infinitely long, then disturbances at frequencies within a wave-passage band can propagate indefinitely, since no reflection occurs at the unit-to-unit interface, whereas those within a wave-stoppage band diminish in a short distance [e.g. 3,4].

However, a structure designed to be periodic can never be exactly periodic. The departure from ideal periodicity is known as disorder. When two adjoining units are not identical, total transmission is no longer possible, and some reflection will take place at the interface. This was first shown by Anderson [5] for an atomic lattice and by Kissel [6] for a generic disordered periodic structure. However, Kissel's analysis requires that the reflection be small; thus, his results are valid only for the case of weak localization.

N. V. Banichuk · D. M. Klimov · W. Schiehlen (Eds.)
Dynamical Problems of Rigid-Elastic
Systems and Structures
IUTAM Symposium Moscow, USSR, May 23-27, 1990

In the present paper, a perturbation procedure is used to compute the so-called localization factor, which is the average decay rate per unit cell of the response amplitude of a disordered periodic structure. This procedure permits successive improvement of the accuracy, making it applicable to either weak, moderate or strong localization. A multiply supported Euler-Bernoulli beam with an additional torsional spring at each support is selected for illustration.

Wave Transmission and Reflection Matrices

Consider two typical cell-units denoted as cells n and n+1. The state vector at the left end of cell-unit n and that at the left end of cell-unit n+1 are related as follows

$$\begin{Bmatrix} w(n+1_\ell) \\ f(n+1_\ell) \end{Bmatrix} = [T(n)] \begin{Bmatrix} w(n_\ell) \\ f(n_\ell) \end{Bmatrix} \tag{1}$$

where each w is a p-dimensional vector of generalized displacements, each f is a p-dimensional vector of generalized forces, and [T(n)] is a 2px2p transfer matrix associated with the nth cell. It is implied that the motion is time-wise sinusoidal, and the displacements and forces are interpreted as their complex-valued amplitudes. Because of disorder, [T(n)] is generally different from the ideal transfer matrix [T] associated with the ideal design condition of exact periodicity. Nevertheless, it is always possible to write

$$[T(n)] = [P(n)][T] \tag{2}$$

which means that the effect of disorder in the nth cell can be lumped at one end and represented by a transfer matrix [P(n)]. Write

$$\begin{Bmatrix} w(n_r) \\ f(n_r) \end{Bmatrix} = [T] \begin{Bmatrix} w(n_\ell) \\ f(n_\ell) \end{Bmatrix}, \qquad \begin{Bmatrix} w(n+1_\ell) \\ f(n+1_\ell) \end{Bmatrix} = [P(n)] \begin{Bmatrix} w(n_r) \\ f(n_r) \end{Bmatrix} \tag{3}$$

Each state vector in (3) can be transformed into a wave vector as follows

$$\begin{Bmatrix} w(n_\ell) \\ f(n_\ell) \end{Bmatrix} = [D]\begin{Bmatrix} \mu^r(n_\ell) \\ \mu^\ell(n_\ell) \end{Bmatrix}, \qquad \begin{Bmatrix} w(n_r) \\ f(n_r) \end{Bmatrix} = [D]\begin{Bmatrix} \mu^r(n_r) \\ \mu^\ell(n_r) \end{Bmatrix} \tag{4}$$

where the columns in [D] are the eigenvectors of [T]. It follows that

$$\begin{Bmatrix} \mu^r(n_r) \\ \mu^\ell(n_r) \end{Bmatrix} = [Q] \begin{Bmatrix} \mu^r(n_\ell) \\ \mu^\ell(n_\ell) \end{Bmatrix} = \begin{bmatrix} \Lambda & 0 \\ 0 & \Lambda^{-1} \end{bmatrix} \begin{Bmatrix} \mu^r(n_\ell) \\ \mu^\ell(n_\ell) \end{Bmatrix} \tag{5}$$

in which Λ is a diagonal matrix of the first p eigenvalues $\lambda_1, \lambda_2, \ldots, \lambda_p$ of the ideal transfer matrix [T], and it can be shown [2] that the other p eigenvalues are λ_1^{-1}, λ_2^{-1}, ..., λ_p^{-1}. Each λ_j represents the ratio of complex-valued amplitudes of one type of wave at the two ends of an ideal cell-unit, which accounts for the reduction of the wave magnitude, as well as the change of phase from one side of an ideal cell-unit to another. Each pair λ_j and λ_j^{-1} corresponds to two identical types of waves: one is right-going and the other is left-going. Eq.(5) can also be re-arranged as follows:

$$\begin{Bmatrix} \mu^\ell(n_\ell) \\ \mu^r(n_r) \end{Bmatrix} = \begin{bmatrix} 0 & \Lambda \\ \Lambda & 0 \end{bmatrix} \begin{Bmatrix} \mu^r(n_\ell) \\ \mu^\ell(n_r) \end{Bmatrix} = [S] \begin{Bmatrix} \mu^r(n_\ell) \\ \mu^\ell(n_r) \end{Bmatrix} \tag{6}$$

The meaning of superscript r (or ℓ) associated with the wave vectors μ is now clear; it indicates the right-going (or left-going) wave groups.

Similar transformation of the state vectors in the second equation in (4) to wave vectors results in

$$\begin{Bmatrix} \mu^r(n+1_\ell) \\ \mu^\ell(n+1_\ell) \end{Bmatrix} = [Q(n)] \begin{Bmatrix} \mu^r(n_r) \\ \mu^\ell(n_r) \end{Bmatrix} = [D]^{-1}[T(n)][D] \begin{bmatrix} \Lambda^{-1} & 0 \\ 0 & \Lambda \end{bmatrix} \begin{Bmatrix} \mu^r(n_r) \\ \mu^\ell(n_r) \end{Bmatrix} \tag{7}$$

Or, re-cast in the form of Eq.(6)

$$\begin{Bmatrix} \mu^\ell(n_r) \\ \mu^r(n+1_\ell) \end{Bmatrix} = [S(n)] \begin{Bmatrix} \mu^r(n_r) \\ \mu^\ell(n+1_\ell) \end{Bmatrix} \tag{8}$$

The elements in [S(n)] can be obtained from those of [Q(n)] as follows

$$s_{11} = -\,q_{22}^{-1}\, q_{21}, \qquad s_{12} = q_{22}^{-1}$$
$$s_{21} = q_{11} - q_{12}\, q_{22}^{-1}\, q_{21}, \qquad s_{22} = q_{12}\, q_{22}^{-1} \tag{9}$$

where s_{ij} and q_{ij} are pxp submatrices of [S(n)] and [Q(n)], respectively, and their argument (n) is omitted for simplicity.

It is meaningful to re-name the submatrices of $[S(n)]$ as $r^r(n)=s_{11}$, $t^\ell(n)=s_{12}$, $t^r(n)=s_{21}$, and $r^\ell(n)=s_{22}$, and re-write Eq.(8) as

$$\begin{Bmatrix} \mu^\ell(n_r) \\ \mu^r(n+1_\ell) \end{Bmatrix} = \begin{bmatrix} r^r(n) & t^\ell(n) \\ t^r(n) & r^\ell(n) \end{bmatrix} \begin{Bmatrix} \mu^r(n_r) \\ \mu^\ell(n+1_\ell) \end{Bmatrix} \qquad (10)$$

The submatrices t and r are called the transmission and reflection matrices, respectively. The superscript for each of these submatrices signifies the direction of an incoming wave group to be transmitted or reflected. For example, $t^r(n)$ is a transmission matrix for an incoming wave group traveling in the right direction.

Matrix $[S(n)]$, which contains the transmission and reflection submatrices, is known as a scattering matrix. This particular scattering matrix characterizes the effects of disorder in cell n which are lumped at the right end of cell n. In wave mechanics, the concept of a scattering matrix is quite general; it characterizes the behavior of any identifiable structural element. In fact, matrix $[S]$ in Eq.(6) is also a scattering matrix, except that the reflection submatries are 0. It is interesting to remark that all scattering matrices are symmetric, a consequence of the reciprocity theorem for linear systems.

For a chain of cell units, 1 through N, we may write

$$\begin{Bmatrix} \mu^r(N+1_\ell) \\ \mu^\ell(N+1_\ell) \end{Bmatrix} = [Q(N,1)]\begin{Bmatrix} \mu^r(1_\ell) \\ \mu^\ell(1_\ell) \end{Bmatrix} = [Q(N)][Q]\ldots[Q(1)][Q]\begin{Bmatrix} \mu^r(1_\ell) \\ \mu^\ell(1_\ell) \end{Bmatrix} \qquad (11)$$

where $[Q]$ and $[Q(n)]$ are given in Eqs. (5) and (7), respectively. Eq.(11) can be re-cast in the form of

$$\begin{Bmatrix} \mu^\ell(1_\ell) \\ \mu^r(N+1_\ell) \end{Bmatrix} = [S(N,1)]\begin{Bmatrix} \mu^r(1_\ell) \\ \mu^\ell(N+1_\ell) \end{Bmatrix} = \begin{bmatrix} r^r(N,1) & t^\ell(N,1) \\ t^r(N,1) & r^\ell(N,1) \end{bmatrix} \begin{Bmatrix} \mu^r(1_\ell) \\ \mu^\ell(N+1_\ell) \end{Bmatrix} \qquad (12)$$

Elements of matrix $[S(N,1)]$ are related to those of matrix $[Q(N,1)]$ according to Eqs.(9). From the second row of Eq.(12), we obtain

$$\mu^r(N+1_\ell) = t^r(N,1)\mu^r(1_\ell) + r^\ell(N,1)\mu^\ell(N+1_\ell) \qquad (13)$$

The Case of Mono-coupling

We shall now consider the case of mono-coupling in more detail, namely, only one type of wave can be transmitted through neighboring cells. In this case the wave transfer matrices Q and the scattering matrices S are 2x2, and the transmission and reflection matrices reduce to the transmission and reflection coefficients. Moreover, we shall assume for the time being that cells N+1, N+2, ..., ∞ are without disorder; thus $\mu^{\ell}(N+1_{\ell})=0$. Then Eq.(13) is simplified to

$$\mu^{r}(N+1_{\ell}) = t^{r}(N,1)\mu^{r}(1_{\ell}) \tag{14}$$

The localization factor is defined as

$$\gamma = - \lim_{\substack{N\to\infty \\ \mu^{\ell}(N+1_{\ell}) = 0}} \frac{1}{N} \ln \left| \frac{\mu^{r}(N+1_{\ell})}{\mu^{r}(1_{\ell})} \right| = - \lim_{N\to\infty} \frac{1}{N} \ln | t^{r}(N,1) | \tag{15}$$

It gives the average exponential decaying rate of the transmitted wave per cell-unit. Several specific cases will now be cited [7]:

(1) Small reflections --- In this case, the effect of multiple reflections is negligible, and

$$\gamma = - \ln | \lambda | + \lim_{N\to\infty} \frac{1}{N} \sum_{n=1}^{N} \ln | q_{22}(n) | \tag{16}$$

Each uncertain parameter may be represented as a mean (or nominal value) plus a random variable. Then $q_{22}(n)$ are functions of such random variables which may be denoted by $q_{22}(n) = q_{22}[X(n)]$ where $X(n) = \{X_1(n),X_2(n),\ldots,X_k(n)\}$ is a vector of random variables with zero means. Assuming that $X_j(n)$ for different n are independent and identically distributed, and that they form an ergodic sequence, the sequential averages in Eq.(16) may be replaced by ensemble averages; namely,

$$\gamma = - \ln | \lambda | + \int \ln | q_{22}(x) | p(x)\, dx \tag{17}$$

where p(x) is the probability density of X(n) and the integrations are carried out in the domain of the random vector X(n).

If $| X(n) | \ll 1$, then Eq.(17) may be approximated by

$$\gamma = - \ln | \lambda | + \frac{\sigma_i^2}{2} \left\{ \frac{\partial^2}{\partial x_i^2} \ln | q_{22}(x) | \right\}_{x = 0} \tag{18}$$

where σ_i^2 are the variances of $X_i(n)$.

(2) Moderate reflections -- If the reflection coefficients are not sufficiently small, then multiple reflections must be considered. By taking reflections from one neighboring cell into account, as illustrated in Fig.1, we obtain

$$\gamma = -\ell n \mid \lambda \mid + \int \ell n \mid q_{22}(x) \mid p(x)dx$$

$$+\iint \ell n \left| 1+ \frac{\lambda^2 \; q_{12}(x_2) q_{21}(x_1)}{q_{22}(x_2) \; q_{22}(x_1)} \right| p(x_1)p(x_2)dx_1dx_2 \tag{19}$$

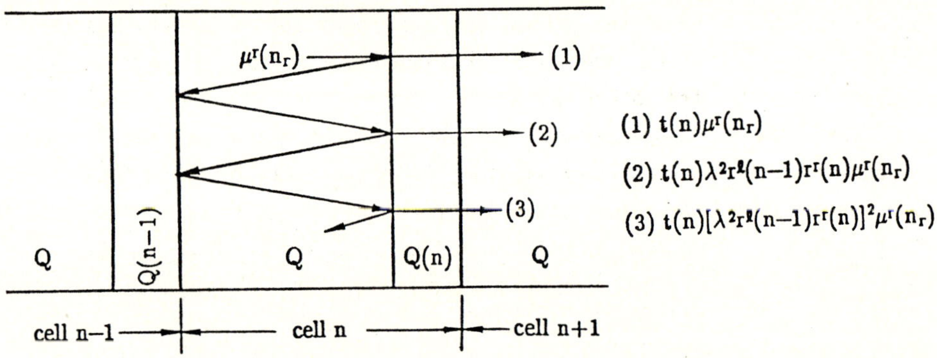

Figure 1 Schematic representation of transmitted waves through the interface betwwen cells n and n+1

(3) Strong reflection -- If the reflection coefficients are relatively large (although each cannot have a magnitude greater than one), it may be necessary to take into consideration the effects of reflection at the interface between cells n-1 and n-2, etc.

An Example

Consider an Euler-Bernoulli beam on evenly spaced hinge supports and with an additional torsional spring at each support. It is assumed that the beam is undamped and that only the torsional spring stiffnesses are random and are described by $k_n = k_0[1+ X(n)]$; $n = 1, 2, \ldots$ where k_0 is the average k_n and $X(n)$ are random variables with zero means. The other physical parameters are assumed to be deterministic, including the distances between neighboring supports ℓ, the bending rigidity of the beam EI, and the mass of the beam per unit length m.

The computed localization factors in the first four wave-passage frequency bands are shown in Fig.2 for δ=0.1, σ=0.1 where $\delta = EI/(k_0\ell)$ and σ is the standard deviation of X(n). Except for frequencies very close to the lower boundary of a wave-passage frequency band all the three equations (17), (18), (19) yield essentially the same results, and in excellent agreement with those obtained from Monte Carlo simulations, also shown in the figure. At the lower boundary of a wave-passage frequency band, the deformations of two neighboring spans tend to be out of phase (exactly out of phase for an ideal periodic structure) and the torsional spring at the supports influence the beam motion to a great extent. Therefore even a small disorder of the torsional springs may give rise to strong localization. On the other hand, the deformations of two neighboring spans are nearly in phase at the upper boundary of a wave-passage frequency and the torsional springs have little effect on the beam motion. Hence, their disorder is relatively unimportant.

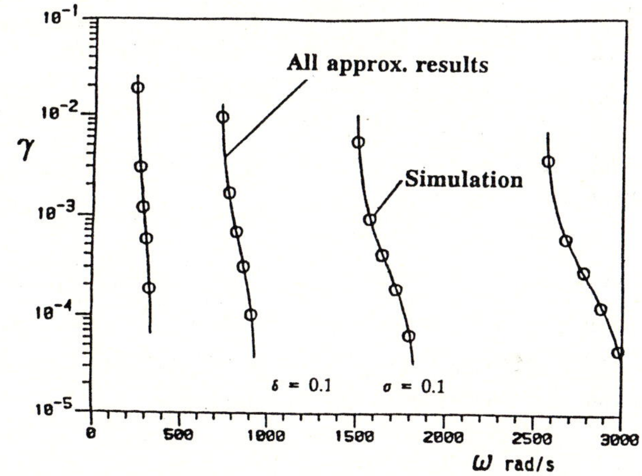

Figure 2 Localization factors at frequences in the first four wave-passage frequency bands

The Monte-Carlo simulations referred to above were carried out by multiplying a large number of random wave transfer matrices at a given frequency to obtain one realization of the wave transfer matrix for the overall N-span system according to Eq.(11) and thus one realization of $t^r(N,1)$. The values of $N^{-1}\ell n\,|\,t^r(N,1)\,|$ for different realizations were averaged over a sufficient number of realizations to obtain an estimate for the localization factor. In our simulations, 300 cells were used in each realization, and the average was taken over 2000 realizations.

Fig.3 depicts the computed localization factors for systems with δ=0.01, σ=0.25 at frequencies in the lower part of the first wave-passage frequency band. It is seen that near the lower boundary of the wave-passage band, Eqs.(17) and (18) overestimate the localization factor, because the effect of multiple reflection is no longer negligible. By taking the reflections in only one cell on the immediate left into account, Eq.(19) yields much more accurate results. This implies that the disorder to internal coupling ratio (σ/δ) is not the only factor in determining the intensity of localization, as once suggested.

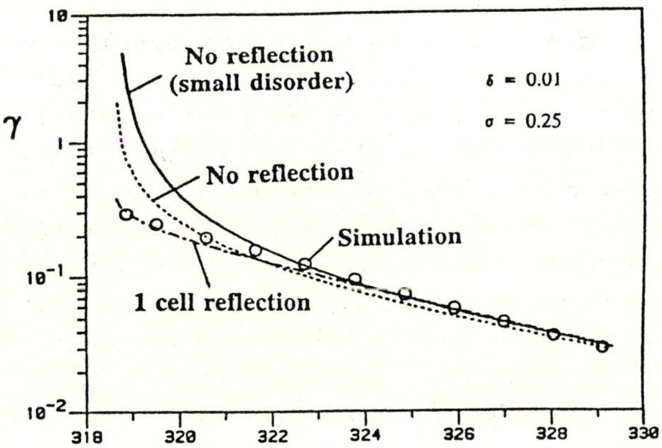

Figure 3 Localization factors at frequences in the lower half of the first wave - passage frequency band

References

1. Miles, J.W., "Vibration of Beams on Many Supports," Journal, Engineering Mechanics Division, ASCE, Vol. 82, EM1, Jan. 1956, pp.1-9.

2. Lin, Y. K. and McDaniel, T. J., "Dynamics of Beam-Type Periodic Structures," Journal of Engineering for Industry, Vol. 91, November, 1969, pp.1133-1141.

3. Mead, D. J., "Wave Propagation and Natural Modes in Periodic Systems: I. Mono-Coupled Systems, II. Multi-Coupled Systems, With and Without Damping," Journal of Sound and Vibration, Vol. 40, 1975, pp.1-39.

4. Yong, Y., and Lin, Y.K., "Propagation of Decaying Waves in Periodic and Piecewise Periodic Structures of Finite Length," Journal of Sound and Vibration, Vol. 129, 1989, pp. 99-118.

5. Anderson, P.W., "Absence of Diffusion in Certain Random Lattices," Physical Review, Vol. 109, 1958, pp. 1492-1505.

6. Kissel, G. J., "Localization in Disordered Periodic Structures," Ph.D. Dissertation, Massachusetts Institute of Technology, 1988.

7. Cai, G. Q. and Lin, Y. K., "Localization of Wave Propagation in Disordered Periodic Structures," to appear in AIAA Journal.

P. Maißer
Academy of Sciences of the G.D.R.
Institute of Mechanics
Chemnitz

Analytical Multi-body Dynamics

1. INTRODUCTION

The classical analytical mechanics goes through an unexpected renaissance caused by rapid developments in spacecraft, robotics, machine and vehicle dynamics as well as in numerical analysis (numerical integration of ordinary differential equations and differential algebraic equations) and symbolic computation (computeralgebra).

A predominant purpuse is the purely computer aided simulation of the dynamical behaviour of Multi Body Systems (MBS) up to the animation. Under computer simulation of the dynamics will be understood the automatic yielding of qualitative and quantitative statements about the dynamical behaviour of a MBS. Qualitative results among others are related to the structure of the model equations, the influence of system parameters on oscillations, the stability of motion including the time-asymptotic and the global behaviour, the existence of first integrals (energy and cyclic integrals) and in case of existence their generation as well as to the properties of periodicity of motion. Quantitative results involve the timedependent behaviour of positions, velocities and accelerations of the MBS, the finding of optimal parameters, optimal control etc.

Lagrange's equations (LE) are a suitable starting point to achieve such statements. An important view point for computer aided generating of other model equations particularly adapted for certain problems is the covariance of LE with respect to kinematic point transformations in the configuration space R^n. The Boltzmann-Hamel equations significant in the nonholonomic mechanics result from a generalization of LE by requiring the covariance with respect to affine transformations in the tangent bundle TR^n of the configuration space R^n. LE can be computer aidedly numerically as well as symbolically generated (NU-/CA-generation) in a derivative-free manner. Both of these methods have specific capabilities, limits, and application fields.

An ingenious coupling of both these methods in dynamics simulation software (hybrid approach) leads to a high efficiency, extactness and safety with respect to the generation of model equations and from them deduced numerical results. The CA generated LE can be investigated by means of CA-methods with respect

N. V. Banichuk · D. M. Klimov · W. Schiehlen (Eds.)
Dynamical Problems of Rigid-Elastic
Systems and Structures
IUTAM Symposium Moscow, USSR, May 23-27, 1990

to the existence of first integrals and simultaneously they are a starting point for an automatic generation of disturbed equations of motion necessary for research on oscillations and stability, see [2, 3].

The algorithm presented in the following due to its algebraic character allows to generate symbolically as well as numerically LE in a uniform wise. Consequently, this algorithm is espacially suited for a flexible hybrid approach in dynamics simulation of MBS.

2. PHENOMENOLOGY OF MULTI-RIGID-BODY SYSTEMS

A MBS is a finite set of rigid bodies embedded in the 3-dimensional euclidean space E^3 which are physically and/or geometrically interconnected together and with a foundation not belonging to the MBS. As bodies are denoted inertial rigid bodies, material points (particles) and noninertial rigid bodies used for geometric interconnections. Physical coupling is specified by applied forces/torques, i.e. by a well-known force relationship of physicallly given forces (spring, damper, gravitational force). Constraints describe geometric interconnections (ideal joints, control motors). All coupling elements are assumed to be inertialess.

3. LAGRANGE'S FORMALISMUS FOR HOLONOMIC MULTI-RIGID-BODY SYSTEMS

The motion of a holonomic scleronomic MBS with n degrees of freedom can be described by the motion of its representing point $q = (q^1,\ldots,q^n)$ in the n-dimensional configuration space R^n. $q^a(a=1,\ldots,n)$ denote the generalized coordinates of the MBS. In the R^n can be introduced a metric g_{ab} (2-fold covariant tensor) defined by the kinetic energy of the MBS $T = {}^1/_2(q)\dot{q}^a\dot{q}^b$, which is a positive definite quadratic form with respect to the generalized velocities $\dot{q}^a$. The configuration space of a MBS R^n so endowed with a Riemannian metric is called a Riemannian space denoted by V^n. $g_{ab} := \dot{\partial}_a\dot{\partial}_b T$ is called metric tensor of V^n (generalized mass matrix), and the kinematical line-element in V^n is defined by $ds^2 = g_{ab}(q)dq^a dq^b$. To the applied forces/torques generalized forces Q_a (n-dimensional covariant vector) are assigned in a well-defined manner. The kinetics of MBS is based on the principle of virtual work in Lagrange's form. Consequently, the motion equations of a MBS are LE of second kind and the dynamics of a MBS may be regarded as a point dynamics in a Riemannian space V^n. All concepts and statements of the Riemannian geometry can be used to describe and to investigate the dynamical behaviour of a MBS.

Starting from the kinetic energy T of a holonomic scleronomic MBS LE

$$(\dot{\partial}_a T)^{\cdot} - \partial_a T = Q_a \tag{1}$$

have the explicit form

$$g_{ab}(q)\ddot{q}^b + \Gamma_{abc}(q)\dot{q}^b\dot{q}^c = Q_a. \tag{2}$$

$$g_{ab} := \sum_{k=1}^{K} [\underset{k}{m}\partial_a \underset{k}{X}^{(i)} \partial_b \underset{k}{X}^{(i)} + \underset{k}{\vartheta}^{ij}\partial_a \underset{k}{E}_i{}^{(i)} \partial_b \underset{k}{E}_j{}^{(i)}] = g_{ba} \tag{3}$$

denote the V^n-metric.

$$\begin{aligned}\Gamma_{abc} &= {}^1/_2(\partial_b g_{ac} + \partial_c g_{ab} - \partial_a g_{bc}) = \\ &= \sum_{k=1}^{K} [\underset{k}{m}\partial_a \underset{k}{X}^{(i)} \partial_b \partial_c \underset{k}{X}^{(i)} + \underset{k}{\vartheta}^{ij}\partial_a \underset{k}{E}^{(i)} \partial_b \partial_c \underset{k}{E}_j{}^{(i)}] = \Gamma_{acb}\end{aligned} \tag{4}$$

are the CHRISTOFFEL symbols of the first kind and

$$Q_a = \sum_{k=1}^{K} [\underset{k}{K}^{(i)} \partial_a \underset{k}{X}^{(i)} + {}^1/_2 \underset{k}{M}^{ij}\partial_a \underset{k}{E}_i{}^{(i)} \underset{k}{E}_j{}^{(i)}] \tag{5}$$

denote the generalized forces, where

$\underset{k}{X}^{(i)}$: cartesian coordinates of the centre of mass of the body B_k,

$\underset{k}{E}_i{}^{(i)}$: matrix of rotation of the body B_k,

$\underset{k}{K}^{(i)}$: components of the resultant of all applied forces acting on the body B_k,

$\underset{k}{M}^{ij}$: two-dimensional skew-symmetric tensor of torque acting on the body B_k,

$\underset{k}{\vartheta}^{ij}$: BINET's inertia tensor of the body B_k,

$k = 1, \ldots, K$;K: number of bodies of a MBS.
Indices in brackets refer to the inertial frame, without brackets to the body frame.

4. MULTI-RIGID-BODY SYSTEMS WITH KINEMATICAL TREE STRUCTURE

Strictly taking into account the special topology, kinematics of determined classes of MBS and using a suitable notation, very important analytical and algebraical properties of functions describing Lagrange's formalism can be discovered.

Assumptions:

(A) topology of MBS: kinematic tree strukture and a regular numeration of the bodies using the "body connection array" ([1]); **(B)** kinematics of MBS: all joints are translational or rational joints (with one degree of freedom) realizing holonomic scleronomic constraints. **(C)** notation: generalized coordinates will be introduced algorithmically by DENAVIT/HARTENBERG notation. **(A)** guarantees the recursive construction of kinematic basic functions (KBF), by which of them metric, CHRISTOFFEL symbols and generalized forces can be generated algebraically. This assumption is not a restriction for constrained mechanical systems (CMS). Cutting open closed loops in CMS always yields MBS with kinematical tree structure. **(B)** requires the equivalent substitution of more general joints (spheric, cardan-joints) by a number of cylindric translational or rotational joints corresponding to the degree of freedom of joint. The use of relative coordinates according **(C)** influence the analytical and algebraical properties of the KBF and consequently of metric as well as CHRISTOFFEL symbols too. All partial derivatives for calculation of metric, CHRISTOFFEL symbols and generalized forces in the traditional calculus due to (3), (4), (5) can be reduced to algebraic operations (exactly!): the kinematics of the MBS in V^n is "algebraized". Defining the KBF ${}_k u_{ia}$, ${}_k\boldsymbol{\Omega}_{ia}$ by the nonholonomic velocities of the bodies B_k

$$ {}_k v_i = {}_k u_{ia}(q)\dot{q}^a, \qquad {}_k\omega_i = {}_k\boldsymbol{\Omega}_{ia}(q)\dot{q}^a, \tag{6} $$

the generalized forces Q_a, the g_{ab} (metric), and the CHRISTOFFEL symbols can be represented as linear, quadratic and cubic forms with respect to them:

$$ Q_a = \sum_{k=a}^{K} [{}_kK^i\,{}_ku_{ia} + (1-s_a)\,{}_kM^i\,{}_k\boldsymbol{\Omega}_{ia}] \tag{7} $$

$$ g_{ab} = \sum_{k=b}^{K} [{}_km\,{}_ku_{ia}\,{}_ku_{ib} + (1-s_a)(1-s_b)\,{}_k\Theta^{ij}\,{}_k\boldsymbol{\Omega}_{ia}\,{}_k\boldsymbol{\Omega}_{jb}]\ ,\quad a \le b, \tag{8} $$

$$ \boldsymbol{\Gamma}_{abc} = (1-s_b) \sum_{k=c}^{K} \varepsilon_i{}^{qr}[{}_km\delta^{ij}\,{}_ku_{ja}\,{}_k\boldsymbol{\Omega}_{qb}\,{}_ku_{rc} + (1-s_a)(1-s_c)\,{}_k\vartheta^{ij}\,{}_k\boldsymbol{\Omega}_{ra}\,{}_k\boldsymbol{\Omega}_{jb}\,{}_k\boldsymbol{\Omega}_{qc}], \tag{9} $$

$a<b\le c$ or $b\le a<c$; ${}_k\Theta^{ij} = {}_k\vartheta_r{}^r\delta^{ij} - {}_k\vartheta^{ij}$.

Generalized forces associated with force coupling elements can be generated in a derivative-free manner too. It is $g_{ab}=g_{ba}$, $\Gamma_{abc}=\Gamma_{acb}$ and $\Gamma_{abc}=-\Gamma_{cba}$ for $b\leq a,c$ due to the especially chosen generalized coordinates (assumption (C)). In view of the symmetry of the metric and of the skew-symmetry of the CHRISTOFFEL symbols of the first kind LE can be generated simultaneously.
The KBF U, Ω can be estimated recursively algebraically:

$$\underset{k}{u}_{ia} = \underset{k}{U}_{ia} + s_a\delta_{3i}\delta_{ka} + (1-s_a)\varepsilon_{si}{}^{j}\underset{k}{h}^{s}\underset{k}{\Omega}_{ja}, \tag{10}$$

$$\underset{k}{U}_{ia} = \begin{cases} [s_a\underset{k}{\hat{E}}_i{}^j + (1-s_a)\underset{k}{d}\varepsilon_{1\mu}{}^{j}\underset{k}{\hat{E}}_i{}^{\mu}]\underset{k}{A}_j{}^3, & a = L(k) \\ \underset{k}{E}_i{}^j[\underset{L(k)}{U}_{ja} + (1-s_a)\varepsilon_{sj}{}^{\varsigma}\underset{k}{Z}^s\underset{L(k)}{\Omega}_{\varsigma a}], & a \in \mathbb{P}_{L^2(k)} \end{cases}, \tag{11}$$

$$\underset{k}{\Omega}_{ia} = (1-s_a)\cdot\begin{cases} \delta_i{}^3, & a=k \\ \underset{k}{E}_i{}^s\underset{L(k)}{\Omega}_{sa}, & a \in \mathbb{P}_{L(k)} \end{cases} \tag{12}$$

Notations: $L(k)$ - body connection array; $s_a(=1,0)$ - distribution parameter of translational and rotational joints,
$q^a=s_a\tau_a+(1-s_a)\varphi_a$ - generalized coordinates,
φ_a, τ_a - complementary joint variables,
$\mathbb{P}_k = \{i|i \in \mathbb{N},\ 1\leq i\leq k,\ i:=L^p(k),\ p=0,1,\dots,\varsigma(k);\ L^{\varsigma+1}(k)=0\}$; $E(\varphi)$- matrix of a planar rotation; A-bifurcation matrix; $E:=\hat{E}A$; d-distance between the joint axes ξ_k and $\xi_{L(k)}$;
$\underset{k}{Z}^s=(\underset{L(k)}{\tau}+6)\underset{k}{\delta}_3{}^s+d\underset{kk}{A}_1{}^s$ - radius-vector $\overrightarrow{O_{L(k)}O_k}$;
$O_{L(k)}$, O_k - "joint points" at $\xi_{L(k)}$, ξ_k respectly, fixed at the body $B_{L(k)}$; $\underset{k}{h}^s=\underset{k}{\tau}\delta_3{}^s+\underset{k}{\xi}^s$ - radius-vector $\overrightarrow{O_kS_k}$; S_k - centre of mass of the body B_k.
The investigation of LE with respect to first integrals (energy/ cyclic integrals) as well as the generation of the disturbed equations of motion

$$g_{ab}\ddot{X}^b + (2\Gamma_{acb}\dot{q}^c-\partial_bQ_a)\dot{X}^b + (\partial_bg_{ac}\ddot{q}^c+\partial_b\Gamma_{acd}\dot{q}^c\dot{q}^d-\partial_bQ_a)X^b = \mu\Phi_a \tag{13}$$

(i.e. the linearized LE about a nominal trajectory) can also be accomplished in a derivative-free manner.

5. CONSTRAINED MECHANICAL SYSTEMS

A scleronomic MBS with kinematical tree structure subject to additional
holonomic (geometrical) constraints: $h^{\varkappa}(q)=0, |\{\varkappa\}|=r_1=\text{rank}(\partial_a h^{\varkappa})$
and/or (14)
nonholonomic constraints: $h^{\lambda}{}_a(q)\dot{q}^a=0, |\{\lambda\}|=r_2=\text{rank}(h^{\lambda}{}_a)$ (15)
with rank $(f^{a_1}{}_a)=r<n$ in a certain domain $D \subset R^n$ and

$$f^{a_1}{}_a = \begin{cases} \partial_a h^{\varkappa}, & a_1=\varkappa \\ h^{\lambda}{}_a, & a_1=\lambda \end{cases}, \quad r=r_1+r_2,$$ is said to be a scleronomic

constrained mechanical system (CMS) with n-r degress of freedom. In case of $r_2=0$ the CMS is said to be holonomic, otherwise nonholonomic. The configuration space of a CMS defined by (14) will be denoted by R^{n-r_1}. With respect to the maximum rank of the matrix $f^{a_1}{}_a$ the following partitioning exists: $(f^{b_1}{}_a)=(f^{b_1}{}_{a_1}/f^{b_1}{}_{a_2})$ with rank $(f^{b_1}{}_{a_1})=r$. Hence, the kinematical constraints $f^{a_1}{}_a(q)\dot{q}^a=0$ can be resolved:

$$\dot{q}^a = F^a{}_{a_2}(q)\dot{q}^{a_2}, \quad q \in R^{n-r_1}, \tag{16}$$

where $F^a{}_{a_2}$ denotes the orthogonal complement of $f^{a_1}{}_a$ with rank $(F^a{}_{a_2})=n-r$: $f^{a_1}{}_a F^a{}_{a_2} \equiv 0$. (16) denotes a parametric representation of a $[2(n-r_1)-r_2]$ - dimensional submanifold of the $2(n-r_1)$ - dimensional tangent bundle $T^{n-r_1}R^{n-r_1}$ (union of all tangent spaces $T^{n-r_1}R^{n-r_1}$ fixed at the points q of the restricted configuration space R^{n-r_1}: $\bigcup_{q\in R^{n-r_1}} T^{n-r_1}R^{n-r_1}$). Operations in view of this submanifold will be denoted by indices endowed with a prime: $\dot{q}^a= F^a{}_{a'_2}(q)\dot{q}^{a'_2}$, $q \in R^{n-r_1}$. The motion equations of a CMS result from LE by projecting into the $[2(n-r_1)-r_2]$ - dimensional tangent bundle $T^{n-r}R^{n-r_1}$ of the virtual displacements $\delta q^a=F^a{}_{a'_2}(q)\delta q^{a'_2}$, $q\in R^{n-r_1}$;

$$g_{a'_2 b'_2}(q)\ddot{q}^{b'_2}+\Lambda_{a'_2 b'_2 c'_2}(q)\dot{q}^{b'_2}\dot{q}^{c'_2} = Q_{a'_2}. \tag{17}$$

The explicit WORONETZ equations (17) have formal the same structure like (2), but they differ from those.

$$g_{a'_2 b'_2} := g_{ab}F^a{}_{a'_2}F^b{}_{b'_2} \tag{18}$$

is together with g_{ab} symmetric, positive definite, and consequently nonsingular, therefore it can be used as a kinematically induced metric in $T^{n-r}R^{n-r_1}$ and finally in the restricted configuration space R^{n-r_1} too. The R^{n-r_1} becomes again a riemannian space V^{n-r_1}, the line-element of which is defined by $ds^2 = g_{a_2' b_2'} dq^{a_2'} dq^{b_2'}$:

$$A_{a_2' b_2' c_2'} := [\Gamma_{abc} - g_{ac_1} \tilde{f}^{c_1}{}_{a_1} \partial_c f^{a_1}{}_b] F^a{}_{a_2'} F^b{}_{b_2'} F^c{}_{c_2'}, \tag{19}$$

$\tilde{f}^{a_1}{}_{b_1}: f^{a_1}{}_{b_1} \tilde{f}^{b_1}{}_{c_1} = \delta^{a_1}{}_{c_1}$, denote the asymmetric (in case of nonholonomicity, i.e. $r_2 \neq 0$) generalized CHRISTOFFEL symbols of the first kind in V^{n-r_1}.
Geometric (holonomic), kinematic, and dynamic (if necessary) constraints can also be generated algebraically using KBF.

6. HYBRID MULTI BODY SYSTEMS

Based on the approach presented in chapter 4 it is also possible to include effects of flexibility in MBS containing elastic substructures due to the ideas created by HUSTON (flexibility and compliance analysis, s. [1]) and RAUH (superelement technique, s. [4]). In both cases LE can be used to obtain the governing equations of motion assuming elastic bodies of the MBS are long slender member which can be modelled as beams with constant cross section. Assuming small elastic deformations the former method for a given finit-segment model of a hybrid MBS can easily be realized for simple geometric structure like slender beams by changing the vector Z^i and the matrix A (s. chapter 4) step by step doing the numerical integration (quasistatic analysis):

$$\underset{r}{Z}^i \longrightarrow \underset{r}{\bar{Z}}^i := \underset{r}{Z}^i + \underset{j}{u}^i, \quad \underset{r}{A}_i{}^s \longrightarrow \underset{r}{\bar{A}}_i{}^s := \underset{r}{A}_i{}^\gamma \underset{r}{B}_\gamma{}^s, \quad j = L(r) \tag{20}$$

$\underset{j}{u}^i$: elastic displacement of the "joint point" O_r,

$\underset{j}{B}_i{}^s$: elastic distorsion of the "joint frame" fixed in O_r, $j=L(r)$,

$$\underset{r}{B}_i{}^s := \delta_i{}^s + \varepsilon_i{}^s{}_\mu \underset{j}{\alpha}^\mu + {}^1/_2 (\underset{j}{\alpha}_i \underset{j}{\alpha}^s - \underset{j}{\alpha}^\gamma \underset{j}{\alpha}_\gamma \delta_i{}^s);$$

"constitutive equations":

$$\underset{j}{u}^i = \underset{j}{K}^i{}_r \underset{j}{f}^r + \underset{j}{L}^i{}_r \underset{j}{m}^r, \quad \underset{j}{\alpha}^i := \underset{j}{\bar{K}}^i{}_r \underset{j}{f}^r + \underset{j}{\bar{L}}^i{}_r \underset{j}{m}^r,$$

$\underset{j}{f}^r$, $\underset{j}{m}^r$: obtained from equilibrium considerations at the undeformed beam including inertia aswell as applied force/torque system;
K, $\bar{K}$, L, $\bar{L}$: obtained by elementary structure analysis.

7. REFERNCES

[1] HUSTON, R.L.: Multi-body dynamics including the effects of Flexibility and Compliance. Computers & Structures, Vol. 14, No. 5-6, 1981, 443-451.

[2] MAISSER, P.: Analytische Dynamik von Mehrkörpersystemen. ZAMM 68(1988)10, 463-481.

[3] MAISSER, P.; HENDEL, K.: Symbolic Computations in Multi-Rigid-Body system Dynamics. 1st Conference on Mechanics, Prague 1987, Proceedings, Volume 6, pp. 198-201.

[4] RAUH, J.: Ein Beitrag zur Modellierung elastischer Balkensysteme. VDI Fortschritt-Berichte, Reihe 18; Mechanik/Bruchmechanik, Düsseldorf, 1987.

A New Approach to the Problem of Modeling for One-dimensional Elastic Systems

F.A. MIKHAJLOV, V.E. MOSOLOV, and L.V. GOUREYEV

Moscow Aviation Institute,
Moscow, USSR

Summary

Mathematical models for the study of the dynamic behaviour of one-dimensional elastic systems are discussed. Models of a new kind especially represented by finite sets of ordinary differential equations of infinite order for the signals on the boundaries are presented. It is shown how to formulate such a model and which advantages it has.

Introduction

Traditionally, mathematical models for the study of the dynamic behaviour of one-dimensional elastic systems are formulated in terms of partial differential equations and boundary conditions. We present here models of other kind especially those represented by the finite sets of ordinary differential equations of infinite order for the signals on the boundaries. In these equations, boundary conditions are taken into account.

Models of system in question in terms of sets of ordinary differential equations are well-known. Such a set is obtained on the base of partial differential equations and contains the infinite number of equations. The models we suggest have the following salient features:

(i) the ordinary differential equations are directly obtained from the given data;

(ii) the complete set of these equations are finite;

(iii) its solutions are sets of the functions describing the signals on the system boundaries;

(iiii) the functions describing the signals inside the system are found by means of recalculation from the above solutions.

N.V. Banichuk · D.M. Klimov · W. Schiehlen (Eds.)
Dynamical Problems of Rigid-Elastic Systems and Structures
IUTAM Symposium Moscow, USSR, May 23-27, 1990

We do not consider here the application of these models for the purpose of analysis. However, we will show that they are convenient for determining such system's characteristics as transfer function and frequency response, since for this purpose it is not required to solve differential equations.

Method of modeling

Consider the one-dimensional elastic system of length m with a horizontal line as spatial coordinate axis and an external signal (force, moment or displacement) at its left end. Let this system be subjected to the linear boundary conditions which are nonhomogeneous for the left end (due to presence of the external signal) and homogeneous for the right one. The block-diagram of this system is shown in Fig.1.

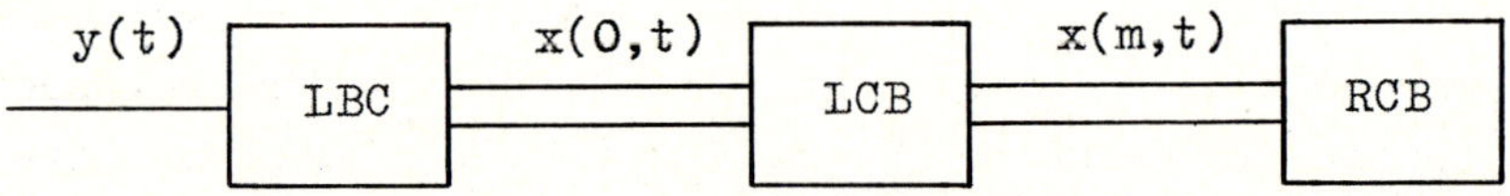

Fig.1. Block-diagram of one-dimensional distributed system

Here LBC is a left boundary-condition block, RBC is a right boundary condition block, and we call LCB a "limit-chain block"; x(0,t) is a k-dimensional vector of signals at the left end; x(m,t) is a k-dimensional vector of signals at the right end, y(t) is an external signal. The limit-chain block represents an elastic system itself,namely,its internal properties determining the relationship between the signals at the left and right ends. In the case of traditional modeling, the mathematical description of LCB gives a set of partial differential equations. We present here an alternative description corresponding more to the term "limit-chain block".

In order to formulate this description, we use the approximation of a distributed system by a lumped system [1,2] . We construct the latter as a chain of n elements where every jth element is a lumped model of the corresponding part of the original system (Fig.2).

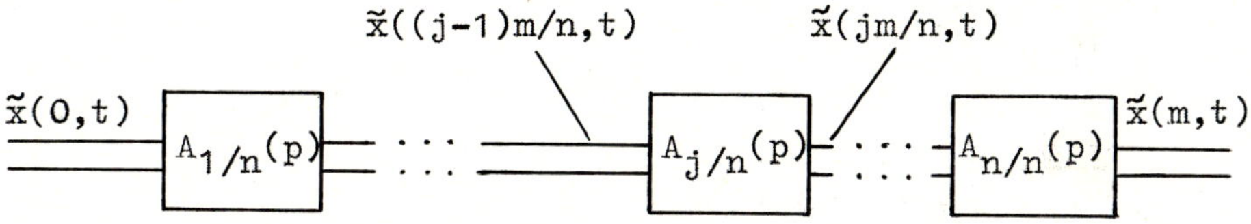

Fig.2. Lumped approximation of one-dimensional distributed system

Here $\tilde{x}(0,t)$ is a vector of signals at the left end of the chain, $\tilde{x}(m,t)$ is a vector of signals at the right end of the chain, $\tilde{x}(jm/n,t)$ is a vector of signals at the right end of jth element, $A_{1/n}(p)$, $A_{j/n}(p)$, $A_{n/n}(p)$ are $k \times k$-matrices whose entries are assumed to be polynomials in $p \equiv d/dt$ with real coefficients depending on m and n. The full set of ordinary differential equations of the motion is as follows

$$\tilde{x}((j-1)m/n,t) = A_{j/n}(p)\tilde{x}(jm/n,t), \; j = 1,\ldots,n. \tag{1}$$

By eliminating from this set the equations containing vector-functions $\tilde{x}(jm/n,t)$, $j = 1,\ldots,n-1$, it is easy to see that

$$\tilde{x}(0,t) = A_{1/n}(p)A_{2/n}(p)\ldots A_{n/n}(p)\tilde{x}(m,t).$$

If $B_n(p) \triangleq A_{1/n}(p)\ldots A_{n/n}(p)$, then

$$\tilde{x}(0,t) = B_n(p)\tilde{x}(m,t). \tag{2}$$

Note that entries $b_{n\mu\nu}(p), \mu, \nu = 1,\ldots,k$, of $B_n(p)$ are polynomials in p, i.e.,

$$b_{n\mu\nu}(p) = b^{(0)}_{n\mu\nu} + b^{(1)}_{n\mu\nu}\,p + \ldots + b^{(\beta_{n\mu\nu})}_{n\mu\nu}p^{n\mu\nu} ,$$

where $\beta_{n\mu\nu}$ is a non-negative integer and $b^{(\alpha)}$, $\alpha = 0,1,\ldots,\beta_{n\mu\nu}$, are real numbers.

Now assume that

$$A(p,0,m) \triangleq \lim_{n\to\infty} B_n(p) \tag{3}$$

exists. Obviously, $A(p,0,m)$ is of the form $\| a_{\mu\nu}(p,0,m) \|_1^k$ where $a_{\mu\nu}(p,0,m)$ are series of p, i.e.,

$$a_{\mu\nu}(p,0,m) = a_{\mu\nu}^{(0)}(0,m)+a_{\mu\nu}^{(1)}(0,m)p + a_{\mu\nu}^{(2)}(0,m)\, p^2+\ldots,$$

and $a_{\mu\nu}^{(\alpha)}(0,m)$ are real numbers.

The concept of existence of $A(p,0,m)$ is considered in the following sense: for every μ, ν and α, there exists the limit

$$\lim_{n\to\infty} b_{\mu\nu}^{(\alpha)}$$

and this limit is equal $a_{\mu\nu}^{(\alpha)}(0,m)$.

It follows from eqns. 2 and 3 that

$$x(0,t) = A(p,0,m)\; x(m,t). \tag{4}$$

This infinite-order ordinary vector differential equation can be looked upon as an alternative description of the limit--chain block.

In order to formulate the description of the full system, we take into account, supplementary to eqn. 4, the equations

$$P_0(p,t)\; x(0,t) + P_y(p,t)y(t) = 0, \tag{5}$$

$$P_m(p,t)x(m,t) = 0, \tag{6}$$

expressing boundary conditions. Here $P_o(p,t)$ and $P_m(p,t)$ are $(k/2)\times k$-matrices whose entries are polynomials in p with real coefficients, in general, t-dependent, and $P_y(p,t)$ is a k/2--dimensional vector whose entries are of the same kind.

Provided the vector functions $x(0,t)$ and $x(m,t)$ and only these ones are of interest, eqns. 4-6 form a complete description of a system. In the more general case where the vector function

$x(l,t)$ for prescribed values of l is of interest, a complete description of system is defined by a set of eqns. 4-6 and equation

$$x(l,t) = A(p,l,m)x(m,t) \tag{7}$$

where $A(p,l,m)$ is a $k \times k$-matrix whose entries are polynomials in p with coefficients depending on m-l similarly to $A(p,0,m)$--entries dependence on m.

Determination of matrix $A(p,0,m)$

In order to determine the entries of matrix $A(p,0,m)$ one may use the procedures described below. Consider at first in more details the structure of $A_{j/n}(p)$-entries. As is mentioned in the previous section, they are polynomials in p. The coefficients of polynomials depend on parameters of jth element. Due to formulation of lumped approximation of system, these parameters may be taken as $\rho_{j1} = r_{j1}(l_j)m/n,\ldots,\rho_{js} = r_{js}(l_j)m/n$ where $r_{j1}(l),\ldots,r_{js}(l)$ are some system parameters per unit length, and l_j is an arbitrary number belonging to interval $[(j-1)m/n, jm/n]$. Moreover, these coefficients themselves are polynomials in m and $r_{j1}(l_j), \ldots, r_{js}(l_j)$ with numerical coefficients dependent on n (see examples in [1] and in the following section).

Now, by taken into account these remarks and referring to matrix $B_n(p)$, we conclude that its entries are also polynomials in p and that the coefficients of these polynomials are polynomials in m and $r_{j\sigma}(l_j)$, $j = 1,\ldots,n$; $\sigma = 1,\ldots,s$. Obviously, the coefficients of the latter polynomials are n-dependent numbers.

It follows from the above mention that the problem considered focuses on the problem of determining the above numerical coefficients. For a given n, the latter problem, in principle, may easily be solved. But, in order to pass from matrix $B_n(p)$ to matrix $A(p,0,m)$, we need in general formulae for generating the coefficients involving positive integer n symbolically. In this version, in solving the problem, difficulties arise.

We point at two ways for obtaining desired formulae, considering in particular the case where the system is homogeneous ($r_{j1},\ldots,$

r_{js} are independent of s) and hence $A_{1/n} \equiv \dots \equiv A_{n/n}$ and $B_n = A^n_{1/n}$.

First, for the case that for a given k and the symbolically represented n, the formulae expressing the entries of matrix M^n as polynomials in the entries of an arbitrary $k \times k$-matix M happen to be known, using these formulae for $M = A_{1/n}$ and expressing the entries of this matrix in terms of m, n and $r_1, \dots, r_s$, p (first index of $r_{j\sigma}$ is omitted due to independence $r_{j\sigma}$ of j) yields the formulae expressing the B_n-entries in terms of the above variables. By regarding these entries as polynomials in p, $m, r_1, \dots, r_s$, we obtain the desired formulae.

Second, by substituting $n = 1,2,\dots,n'$ and expressing B_n-entries as polynomials in $p,m,r_1,\dots,r_s$ with numerical coefficients and by observing the n' sets of these coefficients, form pertinent general formulae for coefficients containing n symbolically and holding for $n \leq n'$. Then reaffirm these formulae for some higher values of n. If negative result happen and n'' is a value of n when this result is cleared up, then we repeat the previous procedure for $n = 1,2,\dots,n''$.

If the formulae for coefficients of B_n-entries as polynomials in p are cleared up, then for every entry, posing $n \to \infty$ and determining the limits yield formulae for coefficients of resulting p-power series representing the related A(p,0,m)-entry.

Example

Consider the small oscillation of a flexible homogeneous beam near its horizontal position. Let the length of the beam m be finite, the ends be free and an external force R(t) be acting in an internal point at a distance l from the left end (Fig.3).

By virtue of hypothesis of linearity, the motion of this beam may be regarded as the superposition of the three motions: the motion due to initial deviation from the horizontal state, the motion due to gravitation forces, and the motion excited by a force R(t). We treat here only the latter motion.

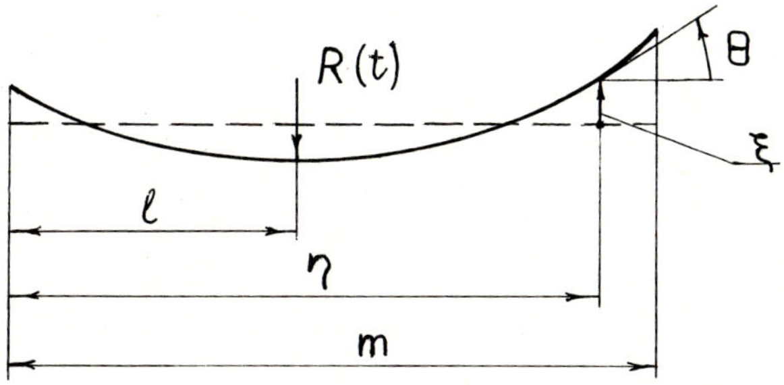

Fig.3. A flexible beam in question

In order to formulate a model, consider a given beam as two beams, one of them (the left one) being of length l and another of length m-l. A block-diagram of a system under consideration is shown on Fig. 4.

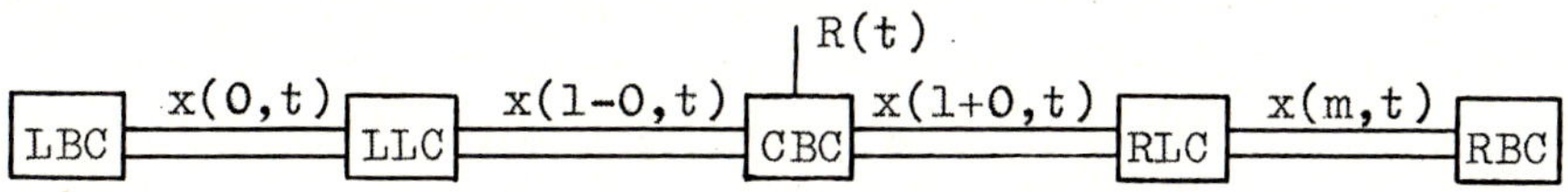

Fig.4. Block-diagram of a beam in question

Here, two limit-chain blocks (left LLC and right RLC) and three boundary-condition blocks (left LBC, right RBC and common CBC) are available. The blocks LLC and RLC represent the left and the right half-beam, respectively. The blocks LBC, RBC and CBC represent boundary conditions at the left end of beam, at the right one at the common point of the two half-beams, respectively. The vector of signals are defined as follows

$$x(\eta,t) \triangleq [M(\eta,t)\ \theta(\eta,t)\ Q(\eta,t)\ \xi(\eta,t)]^T ,$$

where $M(\eta,t)$ is a moment applied at a point η from the right part of the beam to its left part, $Q(\eta,t)$ is a similarly applied force.

Dividing the right half-beam into n equal parts results in the following equation of jth element of its lumped approximation model:

$$\tilde{x}((j-1)v/n,t) = A_{j/n}\tilde{x}(jv/n,t),\ j = 1,\dots,n, \tag{9}$$

where $\tilde{x}(0,t)$ and $\tilde{x}(jv/n,t)$ are lumped-model approximations of $x(l+0,t)$ and $x(\eta_j,t)$ at the points $\eta_j = l+jv/n$, respectively; $v \triangleq m-l$; $A_{j/n} \equiv \| a^{(n)}_{\mu\nu} \|^4_1$ are matrices whose entries are of the form:

$$a^{(n)}_{11} = a^{(n)}_{33} = a^{(n)}_{44} = 1,\ a^{(n)}_{12} = m_b v^3 p^2/4n^3,\ a^{(n)}_{13} = a^{(n)}_{42} = -v^3/n^3,\ a^{(n)}_{14} = a^{(n)}_{32} = -\ m_b v^2 p^2/2n^2,\ a^{(n)}_{21} = v/EIn,$$
$$a^{(n)}_{22} = 1+m_b v^4 p^2/4EIn^4,\ a^{(n)}_{23} = -v^2/EIn^2,\ a^{(n)}_{24} = -m_b v^3 p^2/2EIn^3,$$
$$a^{(n)}_{31} = a^{(n)}_{41} = a^{(n)}_{43} = 0,\ a^{(n)}_{34} = m_b vp^2/n. \tag{10}$$

Here m_b is mass per unit length and I is cross-sectional moment of inertia.

The related equation of the right limit-chain block is of the form

$$x(l+0,t) = A(p,l,m)x(m,t), \tag{11}$$

where $A(p,l,m) = \| a_{\mu\nu} \|^4_1$ is a matrix whose entries are of the form

$$a_{11} = a_{22} = a_{33} = a_{44} = 1 + r^4 v^4 p^2/4! + r^8 v^8 p^4/8! + \dots,$$
$$a_{12} = -EIa_{24} = -EIa_{31} = m_b p^2 EIa_{43} = m_b(v^3 p^2/3! + r^4 v^7 p^4/7! + \dots),\ a_{13} = -EIa_{21} = -a_{34}/m_b p^2 = a_{42} = -(v + r^4 v^5 p^2/5! + \dots),$$
$$a_{14} = m_b p^2 EIa_{23} = a_{32} = m_b p^2 EIa_{41} = -m_b(v^2 p^2/2! + r^4 v^6 p^4/6! + \dots),$$
$$r = \sqrt[4]{m_b/EI}\ . \tag{12}$$

By analogy, the equation of the left limit-chain block is of the form

$$x(l-0,t) \equiv A^{-1}(p,0,l)x(0,t). \tag{13}$$

Since $A^{-1}(p,0,1) \equiv A(p,1,0)$, the entries of this matrix are also p-power series.

The boundary conditions are defined as follows:

$$\begin{aligned} &M(0,t) = M(m,t) = 0, \ M(l-0,t) = M(l+0,t); \\ &\theta(l-0,t) = \theta(l+0,t); \ Q(0,t) = Q(m,t) = 0, \ Q(l-0,t) = \\ &= Q(l+0,t) + R(t); \ \xi(l-0,t) = \xi(l+0,t). \end{aligned} \tag{14}$$

The set of eqns. 11,13,14 provides the complete description of the model in question.

Transfer functions

Return to a set of eqns. 4-6 describing the model of an one-dimensional distributed system. Let t=0 be initial instant, and zero initial conditions let be given for all variables. Let the entries of $P_0(p,t)$, $P_y(p,t)$ and $P_m(p,t)$ be polynomials in p with constant coefficients. Then the second argument can be omitted. Provided the Laplace transforms of all signals exist, applying Laplace transformation and using its properties, we obtain

$$\begin{aligned} &L[x(0,t)] = A(s,0,m)L[x(m,t)], \\ &P_0(s)L[x(m,t)] + P_y(s)L[y(t)] = 0, \\ &P_m(s)L[x(m,t)] = 0. \end{aligned} \tag{15}$$

Regarding y(t) as input and the first entry $x_1(m,t)$ of x(m,t) as output and solving eqns. 15, we obtain

$$L[x_1(m,t)] = W(s)L[y(t)] , \tag{16}$$

where W(s) is transfer function, obviously, defined by A(s,0,m), $P_0(s)$, $P_y(s)$ and $P_m(s)$.

Thus we have obtained an analytical expression for transfer function. Note, it is interesting because the traditional way of transfer function obtaining is linked with solving partial differential equations.

Note, also, that $P_0(s)$, $P_y(s)$ and $P_m(s)$ are polynomials in s and A(s,0,m) is matrix with s-series entries. Hence the complicacy of formulae for W(s) hangs mainly on complicacy of formulae for A(s,0,m)-entries.

In using A(s,O,m), because of series-type entries, problems may arise. However, in the special cases, in which the series can be summarized analytically, they are quite solvable. Particularly in the above example, the matrix A(s,1,m) has the form

$$\begin{bmatrix} S(u) & EIrs^{1/2}V(u) & -T(u)/rs^{1/2} & -EIr^2sU(u) \\ T(u)/EIrs^{1/2} & S(u) & -U(u)/EIr^2s & -rs^{1/2}V(u) \\ -rs^{1/2}V(u) & -EIr^2sU(u) & S(u) & m_bs^{3/2}T(u)/r \\ -U(u)/EIr^2s & -T(u)/rs^{1/2} & rV(u)m_bs^{3/2} & S(u) \end{bmatrix}$$

Here $u = vrs^{1/2}$, $S(u) = (\cosh u + \cos u)/2$, $T(u) = (\sinh u + \sin u)/2$, $U(u) = (\cosh u - \cos u)/2$, $V(u) = (\sinh u - \sin u)/2$. S(u), T(u), U(u) and V(u) are Krylov's functions.

Note, that in the formulae of entries the two-valued power functions $s^{1/2}$ and $s^{3/2}$ participate. For a given s, each of the values $s^{1/2}$ may be taken. The pertinent value of $s^{3/2}$ is defined by the identity $s^{3/2} \equiv ss^{1/2}$.

Conclusions

The key result may be summed up as follows. On the base of lumped approximation of one-dimensional distributed elastic systems a new kind of their models describing in terms of ordinary differential equations of infinite order with time argument can be formulated. Moreover, in each case, the complete set of equations is finite. It is clear how to formulate such a model, some of its advantages are apparent, one of them is the opportunity to obtain in many cases the transfer functions in the analytical form.

References

1. Mikhajlov, F.A. Theory and methods for analysis of time-varying linear systems. Moscow: Nauka 1986, 114-115, 306 (in Russian).
2. Anderson, B.D.O.; Parks, P.C.: Lumped approximation of distributed systems and controllability questions. IRE proceedings 132, N 3 (1985), 89-94.
3. Feodosjev, V.I. Resistance of materials. Moscow: Nauka 1964, 481-484 (in Russian).

Seismically Excited Liquid Waves in Storage Tanks

William A. Nash, R.H. Kirchhoff, and B.Q. Yu

University of Massachusetts, Amherst, MA. USA

Summary

Seismic excitation of an elastic vertical axis cylindrical liquid storage tank leads to finite amplitude waves on the liquid surface. An analysis is presented to indicate wave amplitude corresponding to specified horizontal excitation of the base. Elastic behavior of the tank is assumed and the liquid is taken to be inviscid and incompressible. The analysis is based on an iterative approach involving successive solutions of uncoupled equations for the liquid and the shell wall.

Introduction

Several experimental investigations [1], [2] have indicated that even modest excitations of the base slab of a vertical axis cylindrical liquid storage tank, see Fig. 1, may lead to relatively large amplitude waves on the surface of the liquid. These waves may have amplitudes of the order of one-fourth of the tank radius. In such cases linearized wave theory is no longer valid and it becomes necessary to investigate large amplitude liquid surface motions. To date, there exist several finite element analyses treating this problem. These are due to Liu [3] and Liu and Ma [4] who offered a general mixed Lagrangian-Eulerian approach. At this writing, these programs are not readily available. The objective of the current investigation is to present a hybrid approach involving a finite element representation of the behavior of the elastic cylindrical tank coupled with a perturbation approach to represent the liquid which is taken to be incompressible and inviscid and is represented as a continuum. An alternate approach was offered by Hutton [5] as well as Rogge and Weiss [6] with application to space flight mechanics. The present analysis removes certain restrictions inherent in [5] and [6] and also leads to more rapid convergence of the hybrid approach.

N. V. Banichuk · D. M. Klimov · W. Schiehlen (Eds.)
Dynamical Problems of Rigid-Elastic Systems and Structures
IUTAM Symposium Moscow, USSR, May 23-27, 1990

Analysis

For the thin, elastic cylindrical tank the well-known equations describing linear motion developed by Sanders. Jr. are employed. In these relations the quantities r, θ, and z denote radial, circumferential, and axial coordinates respectively of a point in the shell middle surface. The corresponding displacement components of the tank in the cylindrical coordinate system are denoted by w, v, and u respectively. Only the radial inertial term is retained. The coupled set of three equations in u, v, and w is not given here in the interest of brevity. The boundary conditions of the tank (if clamped at its lower extremity) may be expressed as follows:

$$U = x \cos\omega t$$
$$V = W = 0 \qquad (1)$$
$$\theta_{xy} = \theta_{yz} = \theta_{zx}$$

where U, V, and W are the displacements of the tank at r=R, i.e. at the wall middle-surface.

For a homogeneous, incompressible, inviscid and irrotational liquid in a circular cylindrical tank, the equations of nonlinear motion of the liquid in terms of the velocity potential function Φ are

$$\nabla^2\Phi = 0 \qquad \text{in tank}$$
$$\Phi_t + \frac{1}{2}(\nabla\Phi)^2 + g\eta = 0 \qquad z=\eta+H$$
$$\eta_\tau + \nabla\Phi\ \nabla(z-\eta) = 0 \qquad z=\eta+H \qquad (2)$$
$$\Phi_z = 0 \qquad z=0$$
$$\Phi_r = \varepsilon_0\omega \cos\Theta \cos\omega\tau + \frac{\delta w}{\delta\tau} \qquad r=R$$

If Θ is written as the sum of a disturbance potential function ϕ and a function accounting for the tank motion, i.e.

$$\Phi(r,\theta,z,t) = xr\cos\theta + \phi(r,\theta,z,t) \qquad (3)$$

then under the above assumption the boundary value problem for the liquid can be stated as

$$\nabla^2\phi = 0 \qquad \text{in tank}$$
$$\phi_t + \frac{1}{2}(\nabla\phi)^2 + g\eta = -xr\cos\theta \qquad z=\eta+H$$
$$\eta_\tau + \nabla\phi\ \nabla(z-\eta) = 0 \qquad z=\eta+H \qquad (4)$$
$$\phi_z = 0 \qquad z=0$$
$$\phi_r = \frac{\delta w}{\delta\tau} \qquad r=R$$

The iterative approach adopted here may be designated as the finite element-perturbation method. The first step is to assume that the tank is rigid so that wall boundary conditions are:

$$w_t = 0 \text{ or } \phi_r = 0 \tag{5}$$

The problem of liquid motion may then be solved by the perturbation approach, described later. The two sets of motion equations are uncoupled. Then, the dynamic pressure p(t) of the liquid on the tank wall is obtained. Since p(t) is obtained from a rigid tank, it is only the first approximation.

The second step involves solution of the elastic tank problem by the finite element approach. We assume that p(t) from the first step to be the pressure on the tank wall. This pressure is utilized in the tank equations of motion and through finite elements the displacements and revised boundary conditions are obtained. The boundary condition w_t is approximate since p(t) is approximate.

The third step is to return to the liquid problem and use the approximate boundary condition w_t as the input to the liquid problem. The perturbation method is applied again and a new pressure p(t) is obtained which is an improvement over the value found in the first step.

Then, the new p(t) is used as input into the elastic tank problem, the finite element approach utilized again, and the procedure repeated.

Thus, the coupled boundary-value is uncoupled at every step and the uncoupled problems solved separately. Numerical convergence is found after three of four steps.

For the liquid perturbation method, we assume that the frequency ω of the forced oscillation is close to the lowest natural frequency p_{11} of the tank, since this is the condition under which large amplitude sloshing may arise. In the perturbation approach we adopt the parameter

$$\varepsilon = \frac{\varepsilon_0}{2R} << 1 \tag{6}$$

provided the amplitude ε_0 of excitation is small compared to the tank diameter. The small parameter ε is dimensionless so that coefficients of asymptotic expansion are independent of the system of units.

To express quantitatively the nearness of ω to p_{11}, we introduce a detuning frequency parameter ν defined by

$$\omega^2 = p_{11}^2 (1+\nu\varepsilon^{2/3}) \tag{7}$$

where p_{11} is given by

$$p_{11} = g\lambda_{11} \tanh(\lambda_{11}h) \tag{8}$$

On the basis of the small parameter chosen above, we expand ϕ and η into series in the following form:

$$\phi = \Sigma \phi_n \varepsilon^{n/3}$$

$$\eta = \Sigma \eta_n \varepsilon^{n/3} \tag{9}$$

We shall now simplify the boundary value problem. The dynamic free surface condition and kinematic free surface condition will be expanded in Taylor series about z=H.

First Approximation

The equations for ϕ_1 and η_1 can be written as follows

$$\frac{1}{r}\frac{\delta}{\delta r}\left(r\frac{\delta\phi_1}{\delta r}\right) + \frac{1}{r^2}\frac{\delta^2\phi_1}{\delta\theta^2} + \frac{\delta^2\phi_1}{\delta z^2} = 0 \qquad \text{in tank}$$

$$\phi_{1t} + \frac{\omega^2}{p_{11}^2} g\eta_1 = 0 \qquad z=h$$

$$\phi_{1z} - \eta_{1t} = 0 \qquad z=h \tag{10}$$

$$\phi_{1z} = 0 \qquad z=0$$

$$\phi_{1r} = 0 \qquad r=R$$

The two free surface conditions can be combined to yield

$$\phi_{1tt} + \frac{\omega^2}{p_{11}^2} g\phi_{1z} = 0 \qquad z=H \tag{11}$$

Finally, the elevation of the wave above mean free surface to first order is written as

$$\eta_1 = \frac{p_{11}^2}{\omega g}(\phi_{11}\sin\omega t - \phi_{12}\cos\omega t) \tag{12}$$

Second Approximation

The equations for ϕ_2 and η_2 may be written as:

$$\frac{1}{r}\frac{\delta}{\delta r}\left(r\frac{\delta\phi_2}{\delta r}\right) + \frac{1}{r^2}\frac{\delta^2\phi_2}{\delta\theta^2} + \frac{\delta^2\phi_2}{\delta z^2} = 0$$

$$\phi_{2z} + \frac{\omega^2}{p_{11}^2} g\eta_2 = -(\phi_{1zt}\eta_1 + \phi_{1z}\eta_{1t} + \frac{1}{2}\phi_{1r}^2 + \frac{1}{2r^2}\phi_{1\theta}^2 + \frac{1}{2}\phi_{1z}^2)$$

$$\phi_{2z} - \eta_{2t} = \phi_{1zz}\eta_1 + \phi_{1z}\eta_{1t} - \phi_{1r}\eta_{1r} - \frac{1}{r^2}\phi_{1\theta}\eta_{1\theta} \tag{13}$$

$$\phi_{2z} = 0$$

$$\phi_{2r} = 0$$

and the two liquid free surface equations maybe combined to the form:

$$\phi_{2tt} + \frac{\omega^2}{p_{11}^2} g\phi_{2z} = -[\phi_{1ztt}\eta_1 + 2\phi_{1zt}\eta_{1t} + \phi_{1z}\eta_{1tt} + \phi_{1r}\phi_{1rt} + \frac{1}{r^2}\phi_{1\theta}\phi_{1\theta t}$$

$$+ \phi_{1z}\phi_{1zt} + \frac{\omega^2}{p_{11}^2}(\phi_{1zz}\eta_1 + \phi_{1z}\eta_{1t} - \phi_{1r}\eta_{1r} - \frac{1}{r^2}\phi_{1\theta}\eta_{1\theta})] \tag{14}$$

From the form of Eq. (14) the expressions for ϕ_2, ϕ_{21}, and ϕ_{22} may be written as circular and hyperbolic functions together with Bessel functions. These lengthy forms will not be given here for the sake of brevity. This leads to the second order approximation for height of liquid wave in the form

$$\eta_2 = E_1 + E_2\cos 2\omega t + E_3\sin 2\omega t \tag{15}$$

where the E_1, E_2, and E_3 are tedious functions of ϕ_1, ϕ_2, and η_1. The governing shell equations were solved by a finite element formulation by an implicit direct integration scheme based on a modified Houbolt approach using cubic displacement functions.

Example

The case of a right circular cylindrical shell with axis vertical and of radius 720 inches (18.29 m) wall thickness 1 inch (25.4 mm), made of steel and completely filled with water to a depth of 480 inches (12.12 m) was examined. The base was subjected to the harmonic excitation

$$x = 0.097 \sin(0.01988t) \tag{16}$$

with initial conditions $x_0 = 0$ and $\dot{x}_0 = 0.08647$. Wave heights η and pressures P at various locations in the elastic tank for various times T are shown in Figs. 2, 3, and 4.

Acknowledgement

The authors would like to thank the National Science Foundation for their sponsorship of this work under Grant ECE 8409942. After completion of this work one of the authors (WAN) had the opportunity to visit Prof. K. Schiffner of the Universitat Gesamthochschule, Seigen, Federal Republic of Germany. Independently, and at the same time Prof. Schiffner had developed the same uncoupled, iterative approach to the problem of wave motion in a liquid-filled tank employed in robotics applications.

References

1. Shin, C.F. and Babcock, C.D., "Scale Model Buckling Tests of a Fluid Filled Tank Under Harmonic Excitation", ASME paper No. 80-C2-PVP-66, (1980).

2. Liu, F.J.S., Kirchhoff, R.H., and Nash, W.A., "Tests of Small Scale Liquid Storage Tanks", Recent Advances in Engineering Mechanics and Their Impact on Civil Engineering, Vol. 1, ASCE, (1983), pp. 573-576.

3. Liu, W.K., "Development of Finite Element Procedures for Fluid-Structure Interaction," Ph. D. Dissertation, California Institute of Technology, Pasadena, (1980).

4. Liu, W.K., and Ma, D.C., "Coupling Effect Between Liquid Sloshing and Flexible Fluid-Filled Systems, Nuclear Engineering and Design, Vol. 7, (1982).

5. Hutton, R.E., "An Investigation of Resonant, Nonlinear Nonplaner Free Surface Oscillations of Fluid", NASA TN D-1870, (1963).

6. Rogge, T.R. and Weiss, H.J., "An Approximate Nonlinear Analysis of the Stability of Sloshing Modes Under Translational and Rotational Excitations", NASA CR-220, (1965).

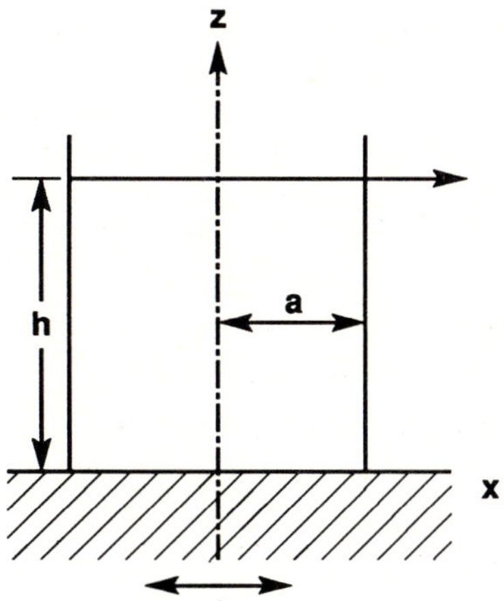

Fig. 1. Nomenclature for cylindrical liquid storage tank

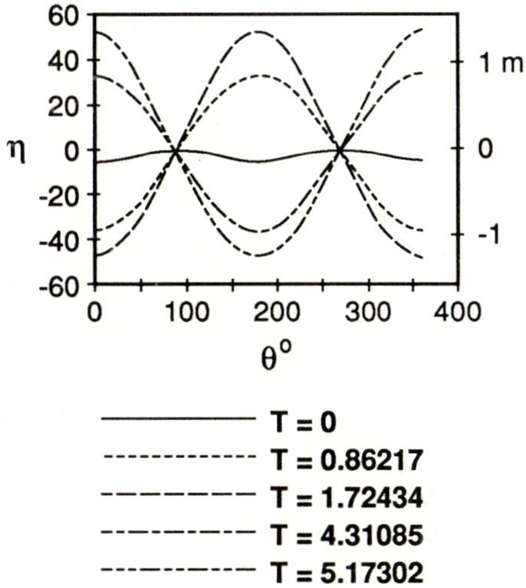

Fig. 2. Variation of wave amplitude with circumference, $\eta - \theta$

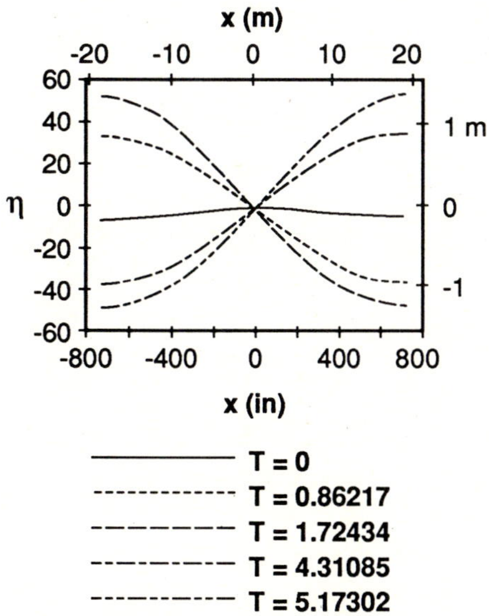

Fig. 3. Distribution of wave amplitude along x-axis, $\eta - x$

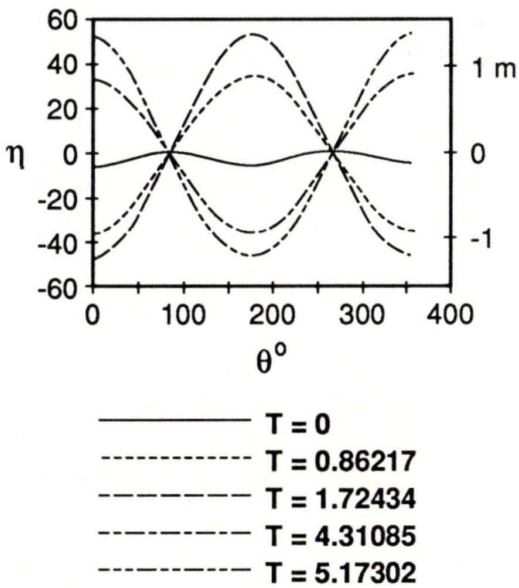

Fig. 4. Variation of wave amplitude with circumference, $\eta - \theta$

Dynamical Simulation of Flexible Multi-body Systems: Applications to Satellites Motions

Madeleine PASCAL

Laboratoire de Modélisation en Mécanique,
associé au CNRS n° 229
Université Pierre et Marie Curie
Tour 66 - 4, place Jussieu
75252 Paris Cédex 05 - France.

Summary

We consider the dynamical analysis of flexible space vehicles modelled by a chain of rigid and elastic bodies with tree structure. The multibody system consists of (n+1) bodies (S_i) (i = 0,1,...,n) interconnected by n hinges 1a (a = 1,...,n). The only external forces and torques are exerted on the first body which is assumed to be rigid. On each individual flexible appendage (S_i), the only external forces and torques are those introduced by the hinges. Assuming that the multibody system undergoes small vibrations around an equilibrium position, we define in frequency domain the linear transformation giving the resultant forces and torques on the boundaries of each flexible appendage (S_i) in terms of the displacements of these boundaries. The motion of each flexible appendage is represented by a set of components modes, which are the modes obtained when the appendage vibrates independently with respect to the other parts of the whole system.

1 - Introduction.

The subject of this work is the dynamical analysis of flexible space vehicules modelled by a chain of rigid and elastic bodies with tree structure. The aim of this study is to obtain the response of the space vehicle to the excitations of external forces and torques, mainly provided by active attitude control.
Typically, the spacecraft consists of a main rigid body (S_o) with a chain of flexible appendages connected to one another. The control system is located only on the main rigid body. The multibody system undergoes small vibrations around an equilibrium position in which the flexible appendages are undeformed.

2 - Kinematics of motion of two contiguous bodies relative to one another.

Let the multibody system consists of n+1 bodies (S_i) (i = 0,1,...,n) interconnected by n hinges 1_a (a = 1,...,n). Each body (S_i) has at most two hinges attached to him (Fig. 1) and each hinge (1_a) is supposed to connect only two bodies named $S_{i^+(a)}$ and $S_{i^-(a)}$ [1]. We assume that on one part $\gamma_{i^+}(a)$ of the boundary of $S_{i^+(a)}$, the elastic displacement is a rigid displacement ; the same assumption is made for one part $\gamma_{i^-(a)}$ of the boundary of $S_{i^-}(a)$. The relative motion of the flexible body $S_{i^-(a)}$ with respect to $S_{i^+(a)}$ is described by the relative motion of the rigid

N.V. Banichuk · D. M. Klimov · W. Schiehlen (Eds.)
Dynamical Problems of Rigid-Elastic
Systems and Structures
IUTAM Symposium Moscow, USSR, May 23-27, 1990

interface $\Upsilon_{i^-(a)}$ with respect to the rigid interface $\Upsilon_{i^+(a)}$: the relative position of $S_{i^+(a)}$ with respect to $S_{i^-(a)}$ depends on N_a degrees of freedom, with $1 \leq Na \leq 6$ according to the kind of articulation between $S_{i^+(a)}$ and $S_{i^-(a)}$.

3 - Problem formulation.

We assume that no distributed external forces act on the flexible appendages. The only external forces acting on each individual appendage (S_i) are the forces introduced by the hinges.
The main rigid body (S_o) is subjected to a set of external actions represented by a force $\vec{F}_o(t)$ in the centre of mass G_o and a torque $\vec{M}_o(t)$.

3.1 - Equations of motion for a flexible appendage.

Let us first consider the motion of one flexible body (S_i), of mass m_i and centre of mass G_i, articulated to two other flexible bodies by means of rigid links (Υ_i) and (Υ'_i) (Fig. 2). Let $(\mathcal{C}_i)$ and $(\mathcal{C}'_i)$ with origins 0_i and $0'_i$ two triads rigidly connected to the rigid interfaces (Υ_i) and (Υ'_i). The flexible body (S_i) undergoes small motions around an equilibrium configuration in which this body is undeformed. Let $(\mathcal{C}_i^o)$ be the position of the triad $(\mathcal{C}_i)$ in the equilibrium configuration. With respect to $(\mathcal{C}_i^o)$, the position of $(\mathcal{C}_i)$ is given by a translation vector $\vec{r_i}$ and a rotation vector $\vec{\alpha_i}$. $\vec{\delta_i}$ is the rotation vector of $(\mathcal{C}'_i)$ with respect to $(\mathcal{C}_i)$.
Let us denotes by $\vec{u_i}(\bar{P},t) = \overline{\bar{P}P}$ the elastic displacement of every material point P of (S_i), $\bar{P}$ being the undeformed position of P, with $\vec{x} = \overrightarrow{0_i\bar{P}}$.
It is assumed that $\vec{r_i}$, $\vec{\alpha_i}$, $\vec{\delta_i}$ are small vectors, of the same order of smallness as the displacement field $\vec{u_i}$.
The linearized equations of motion of the flexible body (S_i) are given by [2] :

$$\left.\begin{aligned}
&L_i\,[u_i] + \rho_i\left(\ddot{r}_i - \tilde{x}\,\ddot{\alpha}_i + \ddot{u}_i\right) = 0 && \text{in } (D_i)\\
&u_i = 0 && \text{on } (\Upsilon_i)\\
&u_i = u_{0'_i} - \left(\tilde{x} - \tilde{x}_{0'_i}\right)\delta_i && \text{on } (\Upsilon'_i)\\
&\sigma_i(u_i)\,\nu_i = 0 && \text{on } \left(\Gamma_{F_i}\right) = \partial D_i - (\Upsilon_i) \cup (\Upsilon'_i)
\end{aligned}\right\} \qquad (1)$$

Here (D_i) is the domain occupied by the flexible body in its undeformed configuration, with boundary ∂D_i, $\vec{\nu_i}$ is the unitarian vector of the outward mormal to ∂D_i, ρ_i is the mass density of (S_i), L_i is a linear self-adjoint differential operator with respect to the components of vector $\vec{x}$ in the $(\mathcal{C}_i)$ frame. $u_{0'_i}$ and $x_{0'_i}$ are the elastic displacement of $0'_i$ and the vector locating the undeformed position of this point in the $(\mathcal{C}_i)$ frame.

σ_i is the stress tensor.
For every vector $\vec{X}$, X denotes the matrix column of its components (X_1, X_2, X_3) in the $(\mathcal{C}_i)$ frame and $\tilde{X}_i$ the shew symmetrie matrix

$$\tilde{X} = \begin{pmatrix} 0 & -X_1 & X_2 \\ X_3 & 0 & -X_2 \\ -X_2 & X_1 & 0 \end{pmatrix}$$

The components in the $(\mathcal{C}_i)$ frame of the resultant force $\overrightarrow{F_i}$ and resultant torque $\overrightarrow{M_i}$ exerted on (Υ_i) are :

$$F_i = \int_{\Upsilon_i} \sigma_i(u_i)\, \nu_i \, dS \qquad M_i = \int_{\Upsilon_i} \widetilde{x_i}\, \sigma_i(u_i)\, \mathcal{C}_i \, dS \Bigg\} \tag{2}$$

Similar expressions are obtained for the force and the torque exerted on (Υ_i') :

$$F_i' = \int_{\Upsilon_i'} \sigma_i(u_i)\, \nu_i \, dS \qquad M_i' = \int_{\Upsilon_i'} \left(\tilde{x}-\tilde{x}_{o_i'}\right) \sigma_i(u_i)\, \nu_i \, dS \Bigg\} \tag{2)'$$

The global motion of (S_i) is deduced from (1), (2) and (2)' :

$$\left.\begin{aligned} m_i\left(\ddot{r}_i - \tilde{l}\,\ddot{\alpha}_i\right) &+ \int_{D_i} \rho_i\, \ddot{u}_i \, dx &= F_i + F_i' \\ m_i\, \tilde{l}_i\, \ddot{r}_i + I_i\, \ddot{\alpha}_i &+ \int_{D_i} \rho_i\, \tilde{x}\, \ddot{u}_i \, dx &= M_i + M_i' + \tilde{x}_{O_i'} F_i' \end{aligned}\right\} \tag{3}$$

Here $\overrightarrow{l_i}$ locates the undeformed position of G_i in $(\mathcal{C}_i)$ and I_i is the inertia tensor in 0_i of the body (S_i) in its undeformed configuration. By use of Fourier's transformation, we obtain from equations (1), (2), (2)', the following equations :

$$\left.\begin{aligned} L_i\left[\bar{v}_i\right] - \rho_i\, \omega^2\, \bar{v}_i &= 0 && \text{in } (D_i) \\ \bar{v}_i &= \bar{r}_i - \tilde{x}\, \bar{\alpha}_i && \text{on } (\Upsilon_i) \\ \overline{v_i} &= \bar{r}_i' - \left(\tilde{x} - x_{o_i'}\right) \bar{\alpha}_i' && \text{on } (\Upsilon_i') \\ \sigma_i\left(\bar{v}_i\right) \nu_i &= 0 && \text{on } \left(\Gamma_{F_i}\right) \end{aligned}\right\} \tag{4}$$

$$\bar{F}_i = \int_{\Upsilon_i} \sigma_i\left(\bar{v}_i\right) \nu_i \, dS \qquad \bar{M}_i = \int_{\Upsilon_i} \tilde{x}\sigma_i\left(\bar{v}_i\right) \nu_i \, dS \tag{5}$$

$$\bar{F}_i' = \int_{\Upsilon_i'} \sigma_i\left(\bar{v}_i\right) \nu_i \, dS \qquad \bar{M}_i' = \int_{\Upsilon_i} \left(\tilde{x} - \tilde{x}_{o_i'}\right) \sigma_i\left(\bar{v}_i\right) \nu_i \, dS \tag{5)'$$

Here $\bar{X}(\omega)$ is the Fourier's transform of $\overrightarrow{X_{(t)}}$,

$$\left.\begin{aligned} \bar{v}_i &= \bar{u}_i + \bar{r}_i - \tilde{x}\,\bar{\alpha}_i \\ \bar{r}'_i &= \bar{u}_{O'_i} + \bar{r}_i - \tilde{x}_{O'_i}\,\bar{\alpha}_i \\ \bar{\alpha}'_i &= \bar{\alpha}_i + \bar{\delta}_i \end{aligned}\right\}$$

3.2 Equations of motion for the main rigid body

The linearized equations of motion of (S_o) are

$$\left.\begin{aligned} m_o\,\ddot{R} &= F_o + F'_o \\ I_o\,\ddot{\theta} &= M_o + \tilde{x}_{o_1}\,F'_o + M'_o \end{aligned}\right\} \quad (6)$$

Here m_o and I_o are the mass and the inertia tensor in G_o of (S_o), R and θ are the column matrices of the $(\mathcal{C}_0)$ components of the vectors $\vec{R}$ and $\vec{\theta}$ giving the inertial position of a triad $(\mathcal{C}_0)$ rigidly connected to (S_0) with respect to an inertial reference frame. This inertial frame is choosen to have R and θ small.
(F_0, M_0) and (F'_0, M'_0) are the $(\mathcal{C}_0)$ components of $(\overrightarrow{F_0}, \overrightarrow{M_0})$, $(\overrightarrow{F'_0}, \overrightarrow{M'_0})$ where to two last vectors are the resultant force and the resultant torque in 0_1 introduced by the link 1_1.
In frequency domain, we obtain :

$$\left.\begin{aligned} - m_0\omega^2\,\bar{R} &= \bar{F}_0 + \bar{F}'_0 \\ - I_0\,\omega^2\,\bar{\theta} &= \bar{M}_0 + \tilde{x}_{0_1}\,\bar{F}'_0 + \bar{M}'_0 \end{aligned}\right\} \quad (6)'$$

4 - Impedance and inertance matrices

The problem formulated by equations (4), (5) and (5)' being a linear problem, we can define a linear transformation depending on the frequency ω giving the resultant forces and torques $\bar{Q}_i = {}^t\left(\bar{F}_i, \bar{M}_i, \bar{F}'_i, \bar{M}'_i\right)$ exerted on the boundaries (γ_i) and (γ'_i) in terms of the displacements $\bar{q}_i = {}^t\left(\bar{r}_i, \bar{\alpha}_i, \bar{r}'_i, \bar{\alpha}'_i\right)$ of these boundaries : $\bar{Q}_i = Z_i(\omega)\,\bar{q}_i$

$Z_i(\omega)$ is a symmetrical matrix of dimensions 12 named the impedance matrix of (S_i). The inverse transformation is given by

$\bar{q}_i = \mathcal{H}_i(\omega)\;\bar{Q}_i$

Here $\mathcal{H}_i(\omega) = [Z_i(\omega)]^{-1}$ is the inertance matrix of (S_i).
For the main rigid body (S_o), the impedance matrix $Z_o(\omega)$ is of dimensions 6 :

$$\left.\begin{aligned} \bar{Q}_0 &= Z_0\ (\omega)\ \bar{q}_0 \\ \bar{Q}_0 &= {}^t\left(\bar{F}_0 + \bar{F}'_0,\ \bar{M}_0 + \bar{M}'_0 + \tilde{x}_{0_1}\ \bar{F}'_0\right) \\ \bar{q}_0 &= {}^t\ (\bar{R},\ \bar{\theta}) \\ Z_0\ (\omega) &= -\ \omega^2 \begin{pmatrix} m_0 E & 0 \\ 0 & I_0 \end{pmatrix} \end{aligned}\right\}$$

E is the identity matrix of order 3.

5 - Component modes

The component modes of a flexible body (S_i) are the modes of vibrations of the body when it vibrates independently with respect to the other parts of the whole system. Two set of component modes can be defined :

5.1 - Contrained modes

Contrained modes are vibrations modes of (S_i) with the rigid links (γ_i) and (γ'_i) fixed $\left(\bar{r}_i = \bar{\alpha}_i = \bar{r}'_i = \bar{\alpha}'_i = 0\right)$. They are defined by :

$$\left.\begin{aligned} & v_i = V_{n_i}(x) \cos\left(\Omega_{n_i} t\right) && n = 1, 2, \ldots \\ & L_i\left[V_{n_i}\right] - \rho_i\ \Omega^2_{n_i}\ V_{n_i} = 0 && \text{in } (D_i) \\ & V_{n_i} = 0 \quad \text{on } (\gamma_i) && \sigma_i\left(V_{n_i}\right) \nu_i = 0 \text{ on } \left(\Gamma_{F_i}\right) \\ & V_{n_i} = 0 \quad \text{on } \left(\gamma'_i\right) && \end{aligned}\right\}$$

We obtain an infinite set of constrained modes, with the orthogonality property :

$$\int_{D_i} \rho_i\ {}^tV_{p_i}\ V_{n_i} dx = 0 \qquad \text{if } \Omega^2_{n_i} - \Omega^2_{p_i} \neq 0$$

The contrained modes can be normalized by setting.

$$\int_{D_i} \rho_i\ {}^tV_{n_i}\ V_{n_i}\ dx = 1 \qquad n = 1, 2, \ldots$$

5-2 Global modes

Global modes are vibrations modes of (S_i) with the rigid links (γ_i) and (γ'_i) free $\left(\bar{F}_i = \bar{M}_i = \bar{F}'_i = \bar{M}'_i = 0\right)$. They are defined by :

$$\left.\begin{aligned}
v_i &= v_{n_i}(x)\cos\left(\omega_{n_i} t\right) && n = 1,2,\ldots \\
L_i\left[v_{n_i}\right] - \rho\omega_{n_i}^2 v_{n_i} &= 0 && \text{in } (D_i) \\
\sigma_i\left(v_{n_i}\right)\nu_i &= 0 && \text{on } \left(\Gamma_{F_i}\right) \\
v_{n_i} &= r_{n_i} - \tilde{x}\alpha_{n_i} && \text{on } (\gamma_i) \\
v_{n_i} &= r'_{n_i} - \left(\tilde{x} - \tilde{x}_{0'_i}\right)\alpha'_{n_i} && \text{on } (\gamma'_i)
\end{aligned}\right\}$$

$$\left.\begin{aligned}
&\int_{\gamma_i}\sigma\left(v_{n_i}\right)\nu_i\, dS = 0 && \int_{\gamma'_i}\sigma\left(v_{n_i}\right)\nu_i\, dS = 0 \\
&\int_{\gamma_i}\tilde{x}\sigma\left(v_{n_i}\right)\nu_i\, dS = 0 && \int_{\gamma'_i}\tilde{x}\sigma\left(v_{n_i}\right)\nu_i\, dS = 0
\end{aligned}\right\}$$

Again, we obtain an infinite set of orthogonal vibrations modes.

6 - Derivation of the impedance/inertance matrices in terms of component modes.

In some particular cases like beams in axial, torsional and bending vibrations, extensible strings or memebranes under uniform tension [3], it is possible to obtain an analytical representation of the matrices $Z_i(\omega)$ and $\mathcal{H}_i(\omega)$. In more general cases, for which a closed form solution for these matrices is not available, we obtain a spectral expension of $Z_i(\omega)$ and $\mathcal{H}_i(\omega)$ in terms of components modes.

6-1 Derivation of the impedance matrix in terms of contrained modes

A spectral expansion of the impedance matrix $Z_i(\omega)$ of each flexible appendage (S_i) is obtained in the form :

$$\left.\begin{aligned}
Z_i(\omega) &= Z_i(0) - \omega^2\mathcal{M}_{b_i} - \omega^4\sum_1^{\infty}\frac{K_{n_i}}{\Omega_{n_i}^2 - \omega^2} \\
\mathcal{M}_{b_1} &= \int_{D_i}\rho_i\ {}^t[\varphi_i]\,[\varphi_i]\,dx \\
K_{n_i} &= \Gamma_{n_i}\,{}^t\Gamma_{n_i} \\
\Gamma_{n_i} &= \int_{D_i}\rho_i\ {}^t[\varphi_i]\,V_{n_i}\,dx
\end{aligned}\right\} \qquad (7)$$

Here $[\varphi_i]$ is the matrix of static deformation modes of (S_i) : $[\varphi_i]$ is a rectangular matrix od dimensions (3 x 12) :

$$[\varphi_i] = [\{\varphi_{i1}\}, \{\varphi_{i2}\}, \{\varphi_{i3}\}, \{\varphi_{i4}\}]$$

$\{\varphi_{ip}\}$ (p = 1,2,3,4) are defined by :

$$\left.\begin{array}{lll} L_i\ [\{\varphi_{ip}\}] & = 0 & \text{in } (D_i) \\ \sigma_i[\{\varphi_{ip}\}]\ \nu_i & = 0 & \text{on } (\Gamma_{Fi}) \\ \{\varphi_{ip}\} & = E\ \delta_{p1} - \tilde{x}\delta_{p2} & \text{on } (\gamma_i') \end{array}\right\}$$

δ_{pK} (p, K = 1, ..., 4) are the Kroneher's symbols.

A convergence property of the matrices K_{ni} (modal gains matrix of the nth constrained mode) holds :

$$\sum_1^{\infty} K_{n_i} = \mathcal{M}_{0_i}$$

This convergence property provides an upper bound of the residual modal gains when only a finite number of terms is used in the spectral expansion (7).

6-2 Derivation of the inertance matrix in terms of global modes.
A spectral expansion of the inertance matrix $\mathcal{H}_i(\omega)$ of each flexible appendage (S_i) is obtained in the form.

$$\left.\begin{array}{lll} \mathcal{H}_i(\omega) & = \mathcal{H}_{i0} - \dfrac{{}^tN_i\ \mathcal{M}_i^{-1}N_i}{\omega^2} + \displaystyle\sum_{n=1}^{\infty} \dfrac{k_{n_i}\ \omega_{n_i}^2\ \omega^2}{\omega_{n_i}^2 - \omega^2} & \\ N_i & = \begin{pmatrix} E & 0 & E & 0 \\ 0 & E & \tilde{x}_{0_i'} & E \end{pmatrix} & \text{rectangular matrix } (6 \times 12) \\ \mathcal{M}_i & = \begin{pmatrix} m_i & -m_i\ \tilde{I}_i \\ m_i\ \tilde{I}_i & I_i \end{pmatrix} & \text{mass matrix in } 0_i \text{ of body } (S_i) \text{ undeformed} \\ k_{n_i} & = \gamma_{ni}\ {}^t\gamma_{ni} & \gamma_{ni} = \displaystyle\int_{D_i} \rho_i\ {}^t[P_i]\ v_{ni}\,dx \end{array}\right\} \quad (8)$$

Here $[P_i] = [\{P_{1i}\}, \{P_{2i}\}, \{P_{3i}\}, \{P_{4i}\}]$ is the matrix of attachment modes defined as static deformation modes of (S_i) subjected to unitarian forces or torques on the boundaries (γ_i) or (γ_i') and corresponding force elements equilibrating these actions.
A convergence property of the matrices k_{ni} (modal gains matrix of the nth global mode) holds :

$$\left.\begin{aligned} \sum_1^{\infty} k_{ni} &= \mathcal{M}_{0i} \\ \mathcal{M}_{0i} &= \int_{D_i} \rho_i \, {}^t[P_i] \; [P_i] dx \end{aligned}\right\}$$

7 - Reduced impedance matrix of the whole system.

For each body (S_i) $(i = 0,1,\ldots,n)$ we obtain an impedance matrix $Z_i(\omega)$ giving the forces and torques exerted on the boundaries in terms of the displacements of these boundaries :

$$\overline{Q_i} = Z_i(\omega)\,\overline{q_i} \qquad i = (0,1,\ldots,n)$$

It results for the whole system the following relation :

$$\left.\begin{aligned} \bar{Q} &= Z(\omega)\bar{q} \\ \bar{Q} &= {}^t\left(\overline{Q_0}, \overline{Q_1}, \ldots, \overline{Q_n}\right) \qquad Z_{(\omega)} = \begin{pmatrix} Z_0 & & 0 \\ & \ddots & \\ 0 & & Z_n \end{pmatrix} \\ \bar{q} &= {}^t\left(\overline{q_0}, \overline{q_1}, \ldots, \overline{q_n}\right) \end{aligned}\right\}$$

$Z(\omega)$ is a symmetrical matrix of order 12n + 6. The set of displacements q is not a set of independant displacements. Each link l_i introduces kinetical linear constraints between the components of q.
We can write the general form of these constraints as : $q = T\,\hat{q}$
Here $\hat{q}$ is the column matrix of a set of independant displacements of the whole system and T is a constant rectangular matrix of maximum rank.
It results for the whole system the relation

$$\left.\begin{aligned} \bar{\hat{Q}} &= \hat{Z}(\omega)\bar{\hat{q}} \\ \hat{Z}(\omega) &= {}^tTZT \qquad \bar{\hat{Q}} = {}^tT\bar{Q} \end{aligned}\right\} \tag{9}$$

$\hat{Z}(\omega)$ is the reduced impedance matrix of the whole system ;
$\hat{Q}$ has the following form : $\bar{\hat{Q}} = \bar{\mathcal{F}} + K(\omega)\bar{\hat{q}}$

Here $\mathcal{F}$ is the column matrix of the external actions for the whole system (exerted only on the main rigid body (S_0)) and $K(\omega)$ is a stiffness matrix associated with the servomotors located in the links.
The final form of the equation (9) is

$$\left.\begin{aligned} H(\omega)\,\bar{\hat{q}} &= \bar{\mathcal{F}} \\ H(\omega) &= Z(\omega) - K(\omega) \end{aligned}\right\} \tag{9'}$$

A modal analysis of the coupled system can be made from (9)' : the equation det $H(\omega) = 0$ gives the frequencies of vibrations of the whole system.

8 - Application to satellites motion.

8.1 Dynamical simulation of a flexible manipulator arm.

We investigate the dynamics of a space manipulator designed to be set on a servicing satellite (S_0) and to grasp another satellite (P). A continuum approach is used to describe distributed flexibility in the manipulator. We obtain for a simplified model the analytical form of the reduced impedance matrix of the whole system.

8.2 Modal analysis and dynamical response of a flexible space station.

We investigate the dynamics of a space station composed of a main rigid body (S_0), two large flexible solar pannels (B_1) and (B_2) and one antenna modelled as a flexible mast rigidly connected to the main body at one end and loaded at the other end. Each solar pannel consists of a rectangular membrane held in tension by a central flexible mast and a rigid tip piece at one end.

The whole system is split up into four substructures, namely the main body, the two solar pannels and the antenna. The reduced impedance matrix of the whole system is then obtained by a spectral expansion in terms of the constrained modes of these substructures.

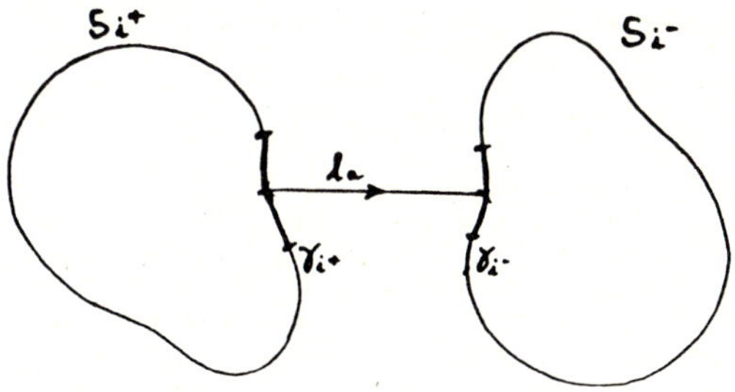

Fig. 1 - Kinematics of two contiguous bodies

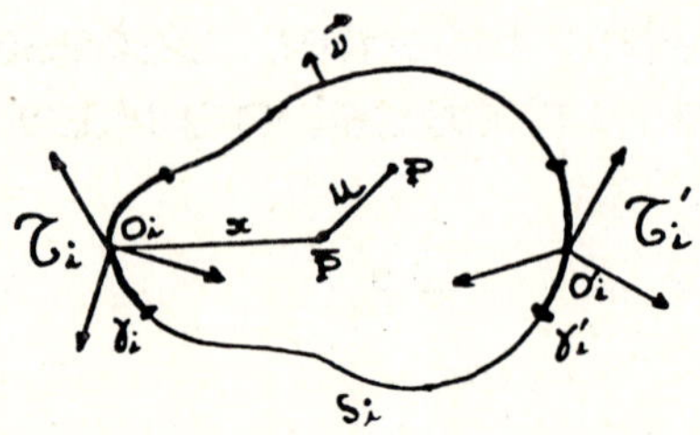

Fig.2 - Dynamics of one flexible body.

References

[1] Wittenburg, J. ; Dynamics of systems of Rigid Bodies. Stuttgart : Teubner 1977

[2] Pascal, M. : Dynamics Analysis of a System of Hinge-connected Flexible Bodies. Celest. Mech. 41 (1988) 253-274.

[3] Kolousek, V. : Dynamics in Engineering Structures. London : Butterworths 1973.

The Geometrical Contact between a Pair of Rollers and a Wheelset in a Railway Vehicle Roller Rig

A.D. DE PATER and YANG GU–ANG

Delft University of Technology

Summary

The theory of the geometrical contact between a railway wheelset and a tangent track, drawn up by the first author two years ago, is extended to the case of a wheelset placed on a pair of rollers in a roller rig. It is shown that in general it is not admissible to extrapolate experimental results obtained by means of a roller rig to the case of a real track. Authors intend to extend the theory to the dynamics of the system track/wheelset and roller rig/wheelset in due time.

1. Introduction

There are two reasons which gave cause for the investigation drawn up in the present paper. The first one is the want to deepen the insight into the geometrical, kinematical and dynamical properties of the roller rig which investigators in the field of railway vehicle system dynamics rather often use. In such a roller rig a vehicle is placed with the wheels on pairs of rollers, which are rotated with a constant angular speed. There are full–scale roller rigs (Munich) but mostly the roller rig and the vehicle are scaled. From the experimental results an investigator can make conclusions about the behaviour of the real vehicle on a real railway track. Recently, Jaschinski has defended a doctor's thesis [1] on this method.

However, the results obtained in this way should be interpreted in a careful way, because the mechanics of the roller rig differ in various aspects essentially from the mechanics of the usual track–vehicle system. Jaschinski did not consider these differences: for him this was admissible, because he restricted himself to rollers with a circular profile of the cross–section and to wheels with purely conical tires with a very small conicity.

Such differences are among other things: the difference in the spin in a roller rig as compared with a tangent track because of the finite value of the roller radius, the difference in the vertical displacement of the wheel–set centre and the influence of the longitudinal displacement of the wheel–set in the case of the roller rig.

N. V. Banichuk · D. M. Klimov · W. Schiehlen (Eds.)
Dynamical Problems of Rigid-Elastic
Systems and Structures
IUTAM Symposium Moscow, USSR, May 23-27, 1990

The second reason for drawing up this paper is the want to obtain a more satisfactory model of a wheelset than the model existing of a wheelset with flangeless wheels and rollers with vertical axes of revolution which limit the lateral motion of the wheelset; such a model was proposed by Bäseler[2]. A better understanding of the contact geometry can also enable us to find a better approximate model of a wheelset than the Bäseler one.

In the present paper we restrict ourselves to the geometry. We build on two previous papers: in [3] we have enunciated the theory of the geometrical contact between a real track and a wheelset and in [4] we have applied this theory numerically on the combination of UIC 60 rails and wheels with the profile S 1002 of the Office of Research and Experiments of the International Railway Union (UIC). The theory can be extended to the (more complicated) system of a roller rig without principal difficulties.

However, we had occasion to apply a principal modification in the way of calculation. In [3], [4] and also previously we always put the origin of the reference co-ordinate system at the height r_0^* (r_0^* being the wheel radius) above the track. But it turned out that it is advantageous to put this origin in the plane of the track: this gives rise to certain simplifications in the equations.

When we consider a wheelset placed on a pair of rollers, the system has four degrees of freedom and there are two constraint relations between the translation co-ordinates u, v, w and the rotatory co-ordinates φ, ϑ, ψ of the wheelset: just as it is the case in the wheelset placed on a tangent track, treated already in [3]. In the latter case the constraint relations contain only the co-ordinates v, w, φ and ψ but in the case of the roller rig also the co-ordinate u appears in these relations. In our present model of the wheelset on a roller rig the case of the tangent track can always be considered as a special case: the case in which the roller radius r_0 is infinite.

As generalized co-ordinates we choose u, φ and ψ, so that v and w are dependent co-ordinates. In the simplified model, however, it turned out that it is more convenient to consider φ as a dependent and v as a generalized co-ordinate.

Because of lack of space many details of the analytical calculations had to be omitted. The interested reader is referred to the Laboratory Report of the same name [5].

At last we draw the attention of the reader to the fact that in [3] some printing errors are present; the formulae for the final results, however, are correct. Moreover, we remark that the symbols r, r_0, and b_0 used in [3], are replaced by r_0^*, $r_y^*-r_y$ and b_0^*.

2. The exact theory

In describing the motion of the rollers and the wheelset we assume the common axis of revolution of the rollers to be horizontal. The asterisk $(*)$ is used to distinguish a wheel quantity from the corresponding roller quantity. In fig. 1 the point M_0 is the centre of the pair of rollers.

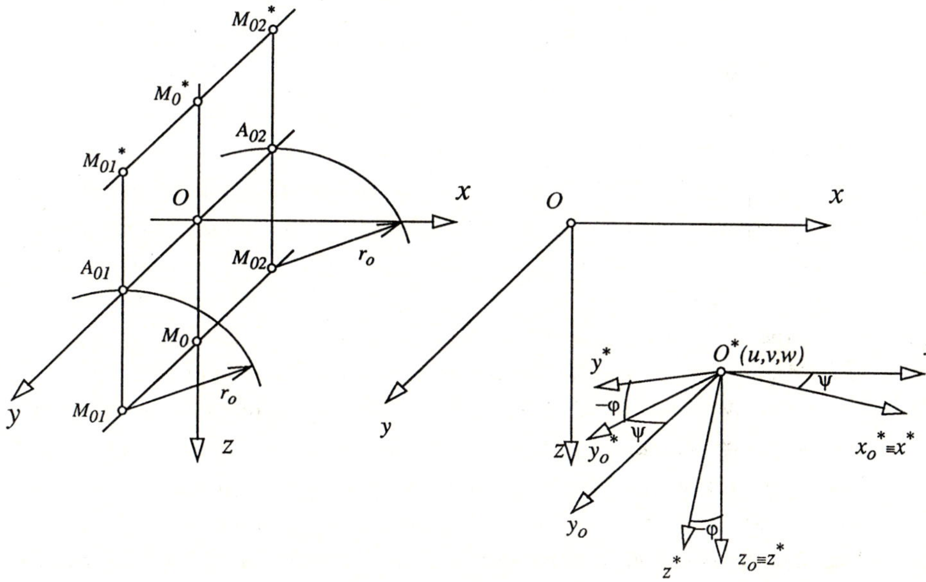

Fig. 1. The motion of the rollers and the wheelset

We first consider the central position of the wheelset. For that position the centre of the wheelset M_0^* is situated on the vertical through M_0. Then the points where a roller contacts a wheel, are A_{0j} (j=1 for the right-hand point and j=2 for the left-hand point). The distance $A_{01}A_{02}$ is designated by $2b$. The verticals through A_{0j} cut the rollers and wheelset axes of revolution in M_{0j} and M_{0j}^*. We call r_0 the distance $A_{0j}M_{0j}$ and r_0^* the distance $A_{0j}M_{0j}^*$.

We now introduce a fixed co-ordinate system (o,x,y,z) in such a way that oy coincides with $A_{01}A_{02}$ and oz with $M_0M_0^*$, where $oA_{01}=oA_{02}$.

The motion of the wheelset is described by the co-ordinate system

(o^*,x^*,y^*,z^*): the origin o^* coincides with the centre of the wheelset and the axis o^*y^* coincides with its axis of revolution. The wheelset can rotate with respect to o^*y^* about the angle ϑ. We can always choose (o^*,x^*,y^*,z^*) in such a way that in the central position of the wheelset it coincides with (o,x,y,z).

The displacement of the wheelset can be described as the sum of the translation with components u, v, w of the wheelset origin o^*, the rotations ψ and φ indicated in fig. 1 and the rotation ϑ of the wheelset about o^*y^*. The latter rotation does not influence the geometrical contact. Now we can find the following relation between the co-ordinates x, y, z and the co-ordinates x^*, y^*, z^*:

$$\begin{bmatrix} x \\ y \\ z \end{bmatrix} = \begin{bmatrix} u \\ v \\ w \end{bmatrix} + \underset{\uparrow \atop G}{\begin{bmatrix} \cos\psi & -\cos\varphi\,\sin\psi & \sin\varphi\,\sin\psi \\ \sin\psi & \cos\varphi\,\cos\psi & -\sin\varphi\,\cos\psi \\ 0 & \sin\varphi & \cos\varphi \end{bmatrix}} \begin{bmatrix} x^* \\ y^* \\ z^* \end{bmatrix}. \tag{2–1}$$

This result enables us to find *six* relations between the co-ordinates x_j,y_j,z_j of the contact points A with respect to the system (o,x,y,z) and the co-ordinates x_j^*,y_j^*,z_j^* of these points with respect to the system (o^*,x^*,y^*,z^*).

More relations can be found from the conditions that the contact point is a point of the surface of a roller and, too, a point of the surface of a wheel; moreover, from the conditions that in a contact point the direction of the normal to the roller surface coincides with the direction of the normal to the wheel surface; see fig. 2a–b.

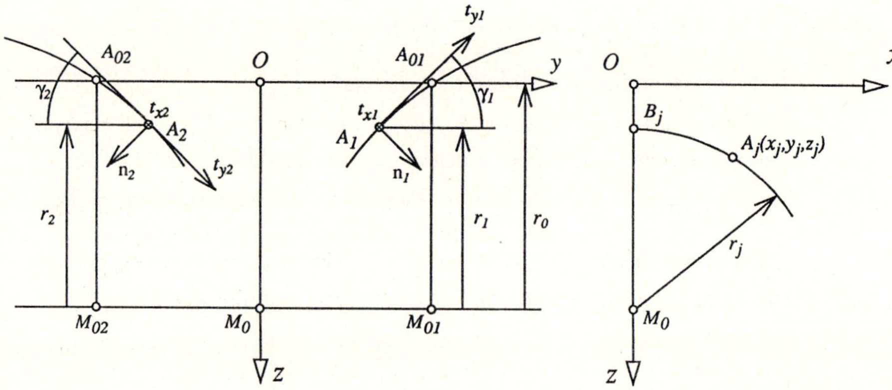

Fig. 2a. Representation of the roller surfaces

The roller and the wheel surface equations are determined by

$$F_j(x_j,y_j,z_j)=\frac{1}{2}\,x_j^2-\frac{1}{2}\,r_j^2+\frac{1}{2}\,(z_j-r_0)^2,$$
$$F_j^*(x_j^*,y_j^*,z_j^*)=\frac{1}{2}\,x_j^{*2}-\frac{1}{2}\,r_j^{*2}+\frac{1}{2}\,(z_j^*+r_0^*)^2. \qquad (2\text{–}2)$$

Here r_j , r_j^* are the distances of the contact point A_j to the axes $M_{01}M_{02}$

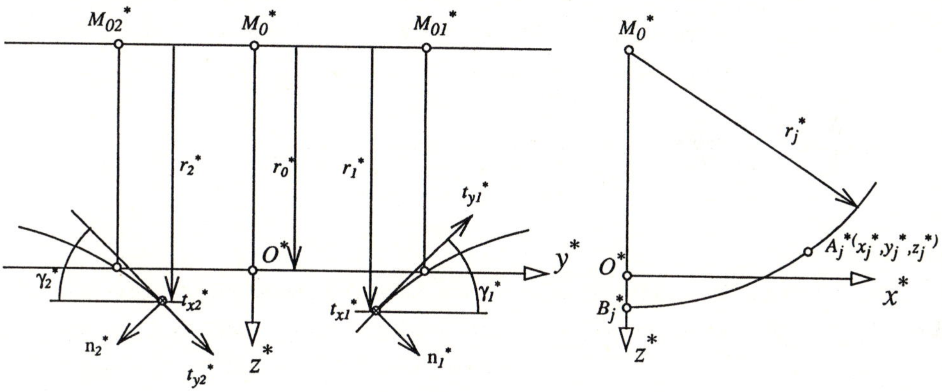

Fig. 2b. Representation of the wheel surfaces

and M_{01}^* M_{02}^* and we can write

$$r_j=f(\pm y_j), \qquad r_j^*=f^*(\pm y_j^*), \qquad (2\text{–}3a\text{–}b)$$

where f, f^* are the *profile functions*, so that $r_0=f(b)$, $r_0^*=f^*(b)$. We also introduce the angles γ_j, γ_j^* in the contact point j:

$$\text{tg }\gamma_j=f'(\pm y_j), \qquad \text{tg }\gamma_j^*=-f^{*\prime}(\pm y_j^*). \qquad (2\text{–}4a\text{–}b)$$

Now the conditions for the contact point to belong to the roller surfa- ce and the wheel surface give rise to the *four* equations

$$F_j(x_j,y_j,z_j)=0, \qquad F_j^*\,(x_j^*,y_j^*,z_j^*)=0. \qquad (2\text{–}5a\text{–}b)$$

For the directions of the normal n in a roller surface with respect to the system (o,x,y,z) and the corresponding directions for the wheel we find

$$n_{xj}=-\frac{x_j\cos\gamma_j}{r_j},\quad n_{yj}=\pm\sin\gamma_j,\quad n_{zj}=-\frac{z_j-r_0}{r_j}\cos\gamma_j, \qquad (2\text{–}6a)$$

$$n_{xj}^*=\frac{x_j^*\cos\gamma_j^*}{r_j^*},\quad n_{yj}^*=\pm\sin\gamma_j^*,\quad n_{zj}^*=\frac{z_j^*+r_0^*}{r_j^*}\cos\gamma_j^*. \qquad (2\text{–}6b)$$

The conditions for the equality of the direction can be expressed by

$$[n^*_{xj},\ n^*_{yj},\ n^*_{zj}]^T = G^{-1}[n_{xj},\ n_{yj},\ n_{zj}]^T : \tag{2-7}$$

again we find six equations but only *four* of them are independent.

Altogether we have 6+4+4=14 equations for the 14 unknowns x_j, y_j, z_j, x^*_j, y^*_j, z^*_j, v and w. Note that r_j and γ_j depend on y_j and r^*_j and γ^*_j on y^*_j: see (2–3a)–(2–4b). From the first set of equations v and w can be eliminated easily and we are left with a set of 12 (non–linear) equations with 12 unknowns. For solving them a computer program has been drawn up.

3. The first–order theory for the roller rig

In railway practice $|x_j|$, $|x^*_j|$, $|z_j|$, $|z^*_j|$ are small, $|y_j|$, $|y^*_j|$ are of the order of b, r_j is of the order of r_0 and r^*_j is of the order of r^*_0Thus we can write

$$x_j=\xi_j,\ y_j=\pm(b-\eta_j),\ z_j=\zeta_j,\ r_j=r_0-\rho_j, \tag{3-1a}$$

$$x^*_j=\xi^*_j,\ y^*_j=\pm(b-\eta^*_j),\ z^*_j=\zeta^*_j,\ r^*_j=r^*_0+\rho^*_j, \tag{3-1b}$$

$$f(\pm y_j) = f(b-\eta_j) = r_0 - \tilde{f}(\eta_j), \tag{3-2a}$$

$$f^*(\pm y^*_j)=f^*(b-\eta^*_j)=r^*_0 + \tilde{f}^*(\eta^*_j). \tag{3-2b}$$

Then also

$$\tilde{\rho}_j=\tilde{f}(\eta_j),\ \mathrm{tg}\ \gamma_j=\left(\frac{d\tilde{f}}{d\eta}\right)_{\eta=\eta_j},\quad \tilde{\rho}^*_j=\tilde{f}^*(\eta^*_j),\ \mathrm{tg}\ \gamma^*_j=\left(\frac{d\tilde{f}^*}{d\eta^*}\right)_{\eta^*=\eta^*_j}. \tag{3-3a–b}$$

Henceforth we shall omit the tilde (˜).

We now consider $|\xi_j|,\dots,|\rho^*_j|$, $|v|$ and $|w|$ as small with respect to b and $|\varphi|$, $|\psi|$ as small with respect to unity and we neglect all terms of order 3 and higher in these quantities. Then the six equations ensuing from (2–1) reduce to

$$\xi_j - \xi^*_j \mp \psi\eta^*_j = u \mp b\psi \tag{3-4a}$$

$$\eta_j - \eta^*_j \pm v \pm \psi\ \xi^*_j \mp \varphi\zeta^*_j = \frac{1}{2}\ b(\varphi^2+\psi^2), \tag{3-4b}$$

$$\zeta_j - \zeta^*_j - w \pm \varphi\ \eta^*_j = \pm b\varphi, \tag{3-4c}$$

the surface equations (2–5) to

$$2r_0(-\rho_j+\zeta_j)=\xi_j^2+\zeta_j^2-\rho_j^2,\qquad 2r^*_0(\rho^*_j-\zeta^*_j)=\xi^{*2}_j+\zeta^{*2}_j-\rho^{*2}_j \tag{3-5a–b}$$

and the equations expressing the equality of the directions to

$$\gamma_j^* = \gamma_j \pm \varphi \pm \frac{\psi\xi_j - \varphi(\zeta_j - \rho_j)}{r_0} - \frac{1}{2}\psi^2 \operatorname{tg}\gamma_j, \tag{3–6a}$$

$$\frac{\xi_j}{r_0} + \frac{\xi_j^*}{r_0^*} \mp \psi \operatorname{tg}\gamma_j + \frac{\xi_j\rho_j}{r_0^2} - \frac{\xi_j^*\rho_j^*}{r_0^{*2}} \mp \frac{\varphi\xi_j^*}{r_0^*} \operatorname{tg}\gamma_j = 0. \tag{3–6b}$$

The equation (3–6a) can be reduced to

$$\gamma_j^* = \gamma_j \pm \varphi \pm \frac{u\ \psi}{r_0 + r_0^*} - \frac{2b + (r_0 - r_0^*)\operatorname{tg}\gamma_j}{2(r_0 + r_0^*)}\psi^2. \tag{3–7}$$

From (3–4b–c) we easily can eliminate v and w and this leads to

$$\eta_1 + \eta_2 - \eta_1^* - \eta_2^* - \varphi(\rho_1^* - \rho_2^*) + \frac{r_0 r_0^*\ (\operatorname{tg}\gamma_1 + \operatorname{tg}\gamma_2)}{r_0 + r_0^*}\psi^2 = b\varphi^2 + \frac{r_0 - r_0^*}{r_0 + r_0^*}b\psi^2, \tag{3–8a}$$

$$\rho_1 - \rho_2 - \rho_1^* + \rho_2^* + \frac{r_0 r_0^*(\operatorname{tg}^2\gamma_1 - \operatorname{tg}^2\gamma_2)}{2(r_0 + r_0^*)}\psi^2 + \varphi(\eta_1^* + \eta_2^*) = 2b\varphi + \frac{2bu\psi}{r_0 + r_0^*}. \tag{3–8b}$$

Thus we are able to reduce the problem to solving *four* (non–linear) equations with four unknowns: η_j and η_j^*; γ_j, γ_j^*, ρ_j and ρ_j^* being functions of η_j and η_j^* because of (3–3a–b).

Having solved these equations we find the constraint equations from (3–4b–c):

$$\mathrm{v} = \frac{1}{2}(-\eta_1 + \eta_2 + \eta_1^* - \eta_2^*) + \frac{r_0^* u\psi}{r_0 + r_0^*} - \frac{r_0 r_0^*(\operatorname{tg}\gamma_1 - \operatorname{tg}\gamma_2)}{2(r_0 + r_0^*)}\psi^2 + \frac{1}{2}\varphi(\rho_1^* + \rho_2^*), \tag{3–9a}$$

$$\mathrm{w} = \frac{1}{2}(\rho_1 + \rho_2 - \rho_1^* - \rho_2^*) + \frac{u^2 + b^2\psi^2}{2(r_0 + r_0^*)} + \frac{r_0 r_0^*(\operatorname{tg}^2\gamma_1 + \operatorname{tg}^2\gamma_2)}{4(r_0 + r_0^*)}\psi^2 + \frac{1}{2}\varphi(\eta_1^* - \eta_2^*). \tag{3–9b}$$

The equations can be solved by means of the Newton–Raphson method. The computer program which for $r_0 = \infty$ has been drawn up for this purpose, is shown in [4], together with a number of numerical results.

4. The completely linear case

When we consider wheelset displacements which are so small that the angles γ_j and γ_j^* do not differ very much from the value γ_0 in the central position, we can write for the profile functions ρ_j, ρ_j^* (3a–b):

$$\rho_j = f(\eta_j) \ = \eta_j \operatorname{tg}\gamma_0 + \eta_j^2 / 2r_y \cos^3\gamma_0\ , \tag{4–1a}$$

$$\rho_j^* = f^*(\eta_j^*) = \eta_j^* \operatorname{tg} \gamma_0 + \eta_j^{*2}/2r_y^* \cos^3\gamma_0, \tag{4-1b}$$

r_y, r_y^* being the roller and wheel profile radii of curvature in the contact points, so that

$$\operatorname{tg} \gamma_j = \operatorname{tg}\gamma_0 + \eta_j/r_y \cos^3\gamma_0,\ \operatorname{tg} \gamma_j^* = \operatorname{tg} \gamma_0 + \eta_j^*/r_y^* \cos^3\gamma_0. \tag{4-2a–b}$$

We now introduce the quantities

$$b_0 = b + r_0 \operatorname{tg} \gamma_0, \quad b_y = b + r_y \sin \gamma_0, \quad \rho_y = \frac{r_y}{r_y^* - r_y}, \tag{4-3a}$$

$$b_0^* = b - r_0^* \operatorname{tg} \gamma_0, \quad b_y^* = b + r_y^* \sin \gamma_0, \quad \rho_y^* = \frac{r_y^*}{r_y^* - r_y}, \tag{4-3b}$$

so that $\rho_y^* - \rho_y = 1$.

We only mention the final results of the calculation for the constraint equations:

$$v = -b\varphi \operatorname{cotg} \gamma_0 - \frac{b_0^* \, u \, \psi \operatorname{cotg} \gamma_0}{r_0 + r_0^*}, \tag{4-4a}$$

$$w = -\frac{1}{2}\{b \operatorname{cotg} \gamma_0 + \frac{b_y b_y^*}{(r_y^* - r_y)\cos \gamma_0 \sin^2\gamma_0}\}\varphi^2 + \frac{u^2}{2(r_0 + r_0^*)} + \frac{b_0 b_0^*}{2(r_0 + r_0^*)}\psi^2. \tag{4-4b}$$

For the tangent track, $r_0 = \infty$ and then

$$v = -b\varphi \operatorname{cotg} \gamma_0, \tag{4-5a}$$

$$w = -\frac{1}{2}\{\operatorname{cotg} \gamma_0 + \frac{b_y b_y^*}{(r_y^* - r_y)\cos \gamma_0 \sin^2\gamma_0}\}\varphi^2 + \frac{1}{2} b_0^* \psi^2 \operatorname{tg} \gamma_0. \tag{4-5b}$$

For the expression between braces we can write

$$c = b \operatorname{cotg} \gamma_0 + \frac{b_y}{\cos \gamma_0 \sin \gamma_0} + \frac{b_y^2}{(r_y^* - r_y)\cos \gamma_0 \sin^2\gamma_0}. \tag{4-6}$$

The last term vanishes for $r_y^* = \infty$. This term is related to the change of γ_j when both the rail and the wheel profile are curved. The first two terms are related to the change of γ_j because of the rotation φ.

The result (4–6) was already found in 1975 by Knothe [6 p. 14] but to his publication not much attention was paid: in the paper [7], in which

Matsui compares results like (4–6) found by various other authors, he makes no mention of Knothe's investigation.

5. The conical wheel on a torus

In the case that the cross-section of the roller is circular and the wheel is purely conical, our problem can be solved in an exact way within the framework of the first order theory; see fig. 3.

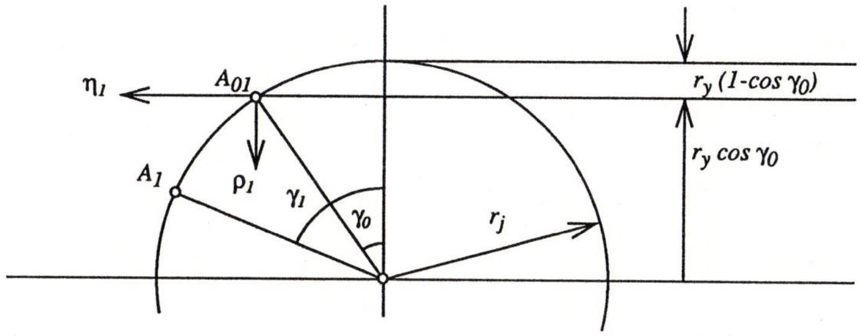

Fig. 3. The right–hand roller with a circular cross–section

In this case we easily find for η_j and ρ_j:

$$\eta_j = r_y(\sin\gamma_j - \sin\gamma_0),\ \rho_j = -r_y(\cos\gamma_j - \cos\gamma_0). \tag{5–1}$$

Moreover, we have for the wheelset:

$$\overset{*}{\gamma}_j = \gamma_0,\ \overset{*}{\rho}_j = \overset{*}{\eta}_j\ \text{tg}\ \gamma_0. \tag{5–2}$$

Now (3–7) yields

$$\gamma_j = \gamma_0 \mp \varphi \mp \frac{u\psi}{r_0 + \overset{*}{r}_0} + \frac{b_0 + \overset{*}{b}_0}{2(r_0 + \overset{*}{r}_0)}\,\psi^2. \tag{5–3}$$

Again, we only mention the final results: we find back the result (4–5a–b) in which, however, the expression between braces now reads

$$c = b\ \text{cotg}\ \gamma + \frac{b_y}{\cos\gamma_0\ \sin\gamma_0}, \tag{5–4}$$

in agreement with (4–6), as it could be expected.

The agreement with the results of section 4 can be understood when we observe that in the present case the conicities again can not differ very much from the values in the central position.

6. Simplified wheelset model for the tangent track

We now come to the second purpose of our investigation, i.e. drawing up a realistic model of a wheelset which can be used for determining the motion along a tangent track.

From numerical calculations for $r_0=\infty$ we have already found that the influence of ψ is rather small. This leads us to omitting all second order terms in (3–6a). Now we even go further and we also omit the term $\pm\varphi$. Moreover, we shall consider v as a generalized co-ordinate rather than φ. In such a way we obtain the equations

$$\xi_j-\overset{*}{\xi}_j=u\mp b\psi,\quad \eta_j-\overset{*}{\eta}_j=\mp v,\quad \zeta_j-\overset{*}{\zeta}_j=w\pm b\varphi, \tag{6–1a–c}$$

$$\zeta_j=\rho_j,\quad \overset{*}{\zeta}_j=\overset{*}{\rho}_j,\quad \overset{*}{\gamma}_j=\gamma_j, \tag{6–2a–c}$$

$$\overset{*}{\xi}_j=\pm\overset{*}{r}_0\ \psi\ \mathrm{tg}\ \gamma_j. \tag{6–3}$$

Now (6–1b) and (6–2c) constitute for each value of j a set of two (non-linear) equations for η_j and $\overset{*}{\eta}_j$, which can be solved for j=1 and j=2 separately. Afterwards we can determine φ and w from

$$b\varphi = \rho_1-\rho_2-\overset{*}{\rho}_1+\overset{*}{\rho}_2,\quad w = \rho_1+\rho_2-\overset{*}{\rho}_1-\overset{*}{\rho}_2 \tag{6–4}$$

and thus we find the constraint relations $\varphi=\varphi(v)$, $w=w(v)$, in which ψ does not appear any more.

We can verify this method by applying it to the completely linear case of sec. 4. As a result we find

$$\varphi = -\frac{v\ \mathrm{tg}\ \gamma_0}{b},\qquad w = -\frac{v^2}{2(\overset{*}{r}_y-r_y)\cos^3\gamma_0}. \tag{6–5}$$

Comparing this with (4–5a–b) we find a complete agreement for φ. In w the agreement for the coefficient c(4–6) of φ^2 is incorrect as far as the influence of the rotation φ is concerned. However, the influence of the curvature does not deviate very much: for the ratio of the two terms at issue we find the value $b_y\overset{*}{b}_y/b^2$, which for small γ_0 goes to 1.

Applying the method to a dicone on rails with a circular cross-section yields again a complete agreement for φ but the value of w now appears to be zero. This could be expected as here the radius $\overset{*}{r}_0$ $=\infty$.

7. Conclusions

We have obtained a consistent theory for the geometrical contact both for a wheelset in a roller rig and for the wheelset on a tangent track. The completely linear system and the Jaschinski roller rig appear as special cases.

For the tangent track we also have drawn up a simplified geometric theory which in future will be used to find a more realistic model of a wheelset on a tangent track than the Bäseler model with vertical rollers. The kinematics and dynamics of such a model will have to be investigated next.

The curve behaviour did not come up for discussion and certainly will have to be considered in such an investigation.

References

1. A. Jaschinski, On the application of similarity laws to a scaled railway bogie model, Diss. Delft University of Technology (1990), 10+158 pp.
2. F. Meineke, Kurzes Lehrbuch des Dampflokomotivbaues, Berlin, Springer (1931) p. 163
3. A.D. de Pater, The Geometrical Contact between Track and Wheelset, Vehicle System Dynamics 17(1988) p. 127–140
4. Yang Gu–ang, A Numerical Procedure for Solving the Non–linear Equations of the Geometrical Contact between Wheelset and Track, Delft University of Technology, Faculty of Mech. Engg. and Marine Techn., Report No. MEMT 7 (1990), 53 pp.
5. A.D. de Pater and Yang Gu–ang, The geometrical contact between a pair of rollers and a wheelset in a railway vehicle roller rig, Delft University of Techn., Lab. for Engg. Mechs. Report No. 910, to be published
6. K. Knothe, Die geometrisch nichtlinearen Beziehungen für einen starren Radsatz, der auf einer starren Schiene querverschoben wird, TU Berlin, Inst. für Luft– und Raumfahrt Report ILR Mitt. 17(1975), 20 pp.
7. N. Matsui, A Re–examination of the Wheel/rail Contact Geometry..., Proc. 7th IAVSD Symposium held at Cambridge, UK, Sept. 7–11, 1981, Lisse (1982) p. 468–480

Elastic Multibody Theory Applied to Elastic Manipulators

F. Pfeiffer, H. Bremer

Institute B of Mechanics, Technical University of Munich, FRG

Summary
We consider multibody systems with rigid and elastic components and the assumption, that any elastic components or couplings generate only small elastic deformations. The multibody system then performs a nonlinear gross motion determined by the corresponding system without elasticities superimposed by small elastic deviations and vibrations. Paper gives the theory for that case and an application on elastic manipulators.

1 Introduction

The research on multibody systems goes on since about thirty years and has created a vast number of scientific contributions and, moreover, a sometimes impetuous discussion about the best way of deriving the equations of motion. Nevertheless, it must be stated, that the fundamental basis of dynamical systems with constraints was established mainly in the 18th and 19th century by the French school of mechanics, d'Alembert, Lagrange [1], [18] and, lateron, in England by Jourdain [13]. The most important finding, namely that passive forces produce neither work nor power, is used in all publications in that area in various forms but always with the same physical result.

The concept of active and passive forces [18], i.e. forces which can be moved or which cannot be moved in their directions, together with the above principles of lost work or lost power lead in any case to the same structure of equations of motion for interconnected rigid bodies, where only evaluation aspects might result in different formal developments [7], [10],[11], [12],[14], [15], [23],[24],[27],[28].

Elastic multibody systems have been treated to a much smaller extend in literature [5], [9], [19], [25], although mechanical systems with elastic components may be found everywhere. Therefore, in the following we shall present a consistent approach to multibody systems with elastic components generating only small elastic deformations. Starting with Jourdain's principle of lost power, lost due to constraints and to the passive character of constraint forces, we establish the equations of motion for a mass element of a rigid or

N. V. Banichuk · D. M. Klimov · W. Schiehlen (Eds.)
Dynamical Problems of Rigid-Elastic
Systems and Structures
IUTAM Symposium Moscow, USSR, May 23-27, 1990

elastic body of the system, we integrate these equations for each body and summarize then for the complete system. The elastic deformations enter through relative kinematics, especially through constraint kinematics, and from there into the corresponding Jacobians and into the momentum equations. The result consists in a set of nonlinear ordinary differential equations containing rigid and elastic generalized coordinates and describing the nonlinear gross motion and the small elastic deviations.

According to this mechanical model the equations of motion may be split up into a set describing the nonlinear gross motion and into a set describing the elastic deformations. Of course, both types of motion are strongly coupled. As is well known, care must be taken in linearizing the elastic motion, because a linear term in the Jacobians multiplied with a large force term in the momentum equations cannot be neglected. As the Jacobians are evaluated by deriving the velocity constraints with respect to the generalized velocities, these constraints must be developed up to quadratic terms in the elastic coordinates.

A series of applications has been worked out in various fields of mechanical engineering, [8],[16],[17],[26]. In the following we shall focus our considerations to elastic manipulators [4],[6],[8],[16],[20], [21],[22].

2 Mechanical and Mathematical Models

2.1 Elastic Manipulators with a Tree-Like Configuration

We start with an elastic robot with a tree-like configuration composed of elastic bar-like links and of elastic joints. Following the properties of motion as discussed above we may describe the complete motion – gross motion and elastic deformations – for each body using three coordinate frames, an inertial one (I in Fig. 2.1), a suitable body-fixed frame B_{oi} and an element-fixed frame B_{Ei}. Note that B_{oi} is already rotated and shifted by the elastic deformations of all bodies with $j < i$. A mass element of the bar is elastically shifted by a vector $\boldsymbol{s}_i(x,t)$ and rotated by a vector $\boldsymbol{\varphi}_i(x,t)$, (Fig. 2.1). The two deformation vectors $\boldsymbol{s}_i$ and $\boldsymbol{\varphi}_i$ are dependent when shear deformation is neglected. With the help of the vectors of Fig. 2.1 the absolute translational and rotational velocities of a mass-element can be written in the form

- B_{oi}-frame:

$$\begin{aligned} \boldsymbol{v}_{Ei} &= \boldsymbol{v}_{oi} + \dot{\boldsymbol{s}}_i + \tilde{\boldsymbol{\omega}}_{oi}(\boldsymbol{x}_i + \boldsymbol{s}_i) \\ \boldsymbol{\omega}_{Ei} &= \boldsymbol{\omega}_{oi} + \dot{\boldsymbol{\varphi}}_i, \end{aligned} \tag{2.1}$$

- B_{Ei}-frame:

$$\begin{aligned} \boldsymbol{v}_i &= (\boldsymbol{E} + \tilde{\boldsymbol{\varphi}}_i)^T \boldsymbol{v}_{Ei}, \\ \boldsymbol{\omega}_i &= (\boldsymbol{E} + \tilde{\boldsymbol{\varphi}}_i)^T \boldsymbol{\omega}_{Ei}. \end{aligned} \tag{2.2}$$

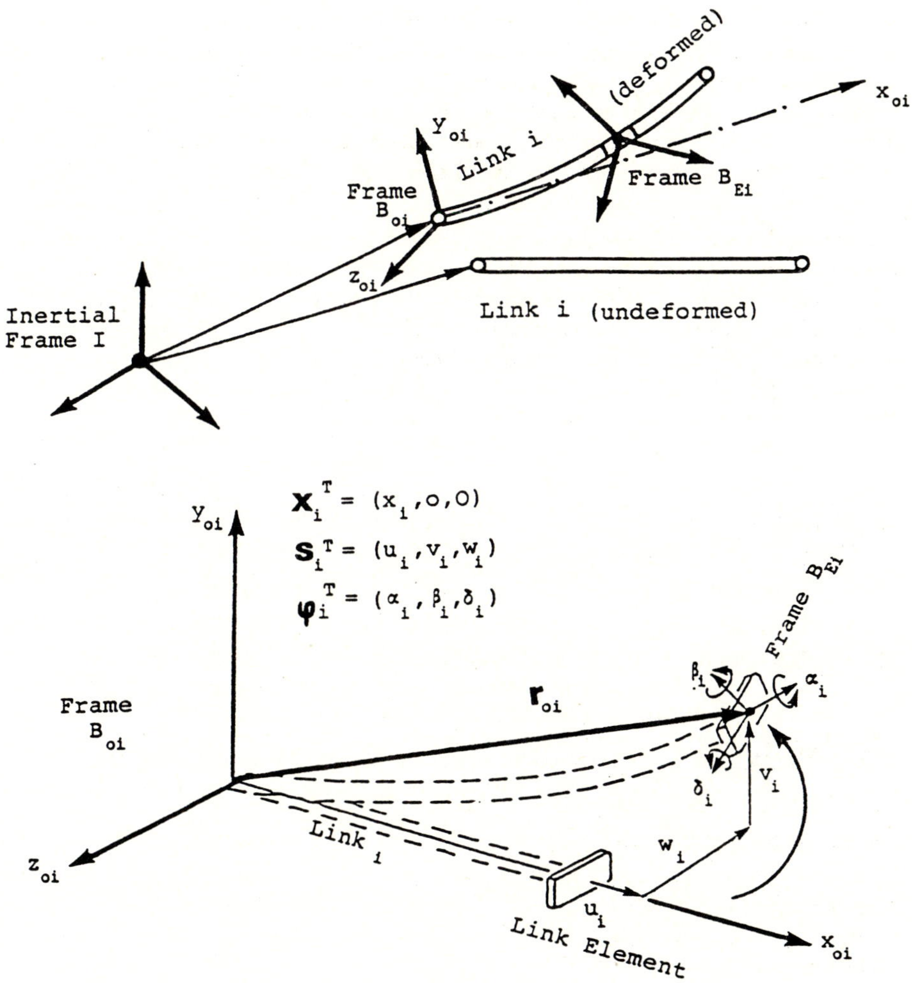

Fig. 2.1: Coordinate Frames and Details of Link i

The absolute accelerations in the two frames are consequently:

- B_{oi}-frame:

$$\begin{aligned} \dot{\boldsymbol{v}}_{Ei} &= \dot{\boldsymbol{v}}_{oi} + \dot{\tilde{\boldsymbol{\omega}}}_{oi}(\boldsymbol{x}_i + \boldsymbol{s}_i) + \tilde{\boldsymbol{\omega}}_{oi}\tilde{\boldsymbol{\omega}}_{oi}(\boldsymbol{x}_i + \boldsymbol{s}_i) + 2\tilde{\boldsymbol{\omega}}_{oi}\dot{\boldsymbol{s}}_i + \ddot{\boldsymbol{s}}_i, \\ \dot{\boldsymbol{\omega}}_{Ei} &= \dot{\boldsymbol{\omega}}_{oi} + \ddot{\boldsymbol{\varphi}} \end{aligned} \tag{2.3}$$

- B_{Ei}-frame:

$$\dot{\boldsymbol{\omega}}_i = (\boldsymbol{E} + \tilde{\boldsymbol{\varphi}}_i)^T \dot{\boldsymbol{\omega}}_{Ei} + \dot{\tilde{\boldsymbol{\varphi}}}^T \boldsymbol{\omega}_{Ei}. \tag{2.4}$$

It is obvious, that all magnitudes in these equations depend on rigid and elastic generalized coordinates $\boldsymbol{q}_r \in \mathbb{R}^{f_r}$, $\boldsymbol{q}_e \in \mathbb{R}^{f_e}$. The elastic coordinates enter the kinematical equations via a Ritz approach, which for $\boldsymbol{s}_i$ can be written in the form

$$\boldsymbol{s}_i(\boldsymbol{q}_e, \boldsymbol{x}_i) = \begin{pmatrix} u_i \\ v_i \\ w_i \end{pmatrix} = \begin{pmatrix} \boldsymbol{q}_{e,u}^T(t)\boldsymbol{u}(\boldsymbol{x}) \\ \boldsymbol{q}_{e,v}^T(t)\boldsymbol{v}(\boldsymbol{x}) \\ \boldsymbol{q}_{e,w}^T(t)\boldsymbol{w}(\boldsymbol{x}) \end{pmatrix}_i . \tag{2.5}$$

Applying this approach, collecting all elastic coordinates in a vector $\boldsymbol{q}_e$ and taking into account the dependencies $\boldsymbol{v}_{Ei} = \boldsymbol{v}_{Ei}(\boldsymbol{q}_r, \boldsymbol{q}_e, \dot{\boldsymbol{q}}_r, \dot{\boldsymbol{q}}_e)$, $\boldsymbol{\omega}_i = \boldsymbol{\omega}_i(\boldsymbol{q}_r, \boldsymbol{q}_e, \dot{\boldsymbol{q}}_r, \dot{\boldsymbol{q}}_e)$ results in

$$\begin{aligned} \boldsymbol{v}_{Ei} &= \left[\left(\frac{\partial \boldsymbol{v}_{oi}}{\partial \dot{\boldsymbol{q}}_r}\right) + (\tilde{\boldsymbol{x}}_i + \tilde{\boldsymbol{s}}_i)^T \left(\frac{\partial \boldsymbol{\omega}_{oi}}{\partial \dot{\boldsymbol{q}}_r}\right)\right] \dot{\boldsymbol{q}}_r + \\ & \quad \left[\left(\frac{\partial \boldsymbol{v}_{oi}}{\partial \dot{\boldsymbol{q}}_e}\right) + (\tilde{\boldsymbol{x}}_i + \tilde{\boldsymbol{s}}_i)^T \left(\frac{\partial \boldsymbol{\omega}_{oi}}{\partial \dot{\boldsymbol{q}}_e}\right) + \left(\frac{\partial \dot{\boldsymbol{s}}_i}{\partial \dot{\boldsymbol{q}}_e}\right)\right] \dot{\boldsymbol{q}}_e, \\ \boldsymbol{\omega}_i &= \left(\frac{\partial \boldsymbol{\omega}_{oi}}{\partial \dot{\boldsymbol{q}}_r}\right) \dot{\boldsymbol{q}}_r + \left[\left(\frac{\partial \boldsymbol{\omega}_{oi}}{\partial \dot{\boldsymbol{q}}_e}\right) + (\boldsymbol{E} + \tilde{\boldsymbol{\varphi}}_i)^T \left(\frac{\partial \boldsymbol{\varphi}_i}{\partial \dot{\boldsymbol{q}}_e}\right)\right] \dot{\boldsymbol{q}}_e, \\ \boldsymbol{q}_r &\in \mathbb{R}^{f_r}, \ \boldsymbol{q}_e \in \mathbb{R}^{f_e}, \ \left(\frac{\partial \boldsymbol{v}_{oi}}{\partial \dot{\boldsymbol{q}}_r}\right) \in \mathbb{R}^{3,f_r}, \ \left(\frac{\partial \dot{\boldsymbol{s}}_i}{\partial \dot{\boldsymbol{q}}_e}\right) \in \mathbb{R}^{3,f_e}, \text{etc.} \end{aligned} \tag{2.6}$$

Jourdain's principle says that passive forces generate no power. Passive forces are forces which cannot be displaced. Conversely, active forces can be displaced. If we apply the momentum equation to the mass-element of Fig. 2.1, arrange the forces with respect to active and passive properties and use Jourdain's principle, we come out with

$$\sum_{i=1}^{n} \int_{B_i} \Delta \boldsymbol{v}_{Ei}^T \left[(\dot{\boldsymbol{v}}_{Ei} + \tilde{\boldsymbol{\omega}}_{oi} \boldsymbol{v}_{Ei})\, dm_i - d\boldsymbol{F}_{ai} \right] = 0, \tag{2.7}$$

where the applied forces $\boldsymbol{F}_{ai}$ contain the external applied forces $\boldsymbol{f}_i$ and the elastic forces $(div\boldsymbol{S})/\rho$ with the strain tensor $\boldsymbol{S}$:

$$\boldsymbol{F}_{ai} = \frac{1}{\rho_i} div\boldsymbol{S}_i + \boldsymbol{f}_i \tag{2.8}$$

The strain tensor $\boldsymbol{S}_i$ must be evaluated from the deformation vectors $\boldsymbol{s}_i$ and $\boldsymbol{\varphi}_i$, Fig. 2.1, for each special configuration. According to the assumption of small elastic deformations we apply Hooke's law. Combining eq.(2.6) to eq.(2.8) we get the following power budget

$$\sum_{i=1}^{n} \int_{B_i} \left\{ \left[\left(\frac{\partial \boldsymbol{v}_{oi}}{\partial \dot{\boldsymbol{q}}_r} \right) + (\tilde{\boldsymbol{x}}_i + \tilde{\boldsymbol{s}}_i)^T \left(\frac{\partial \boldsymbol{\omega}_{oi}}{\partial \dot{\boldsymbol{q}}_r} \right) \right] \Delta \dot{\boldsymbol{q}}_r + \right.$$

$$\left. \left[\left(\frac{\partial \boldsymbol{v}_{oi}}{\partial \dot{\boldsymbol{q}}_e} \right) + (\tilde{\boldsymbol{x}}_i + \tilde{\boldsymbol{s}}_i)^T \left(\frac{\partial \boldsymbol{\omega}_{oi}}{\partial \dot{\boldsymbol{q}}_e} \right) + \left(\frac{\partial \dot{\boldsymbol{s}}_i}{\partial \dot{\boldsymbol{q}}_e} \right) \right] \Delta \dot{\boldsymbol{q}}_e \right\} \times$$

$$\times \left[(\dot{\boldsymbol{v}}_{Ei} + \tilde{\boldsymbol{\omega}}_{oi} \boldsymbol{v}_{Ei}) - \boldsymbol{f}_i - \frac{1}{\rho_i} div\boldsymbol{S}_i \right] dm_i = 0, \tag{2.9}$$

which can be fulfilled by setting the parts connected with $\Delta\dot{\boldsymbol{q}}_r$ and $\Delta\dot{\boldsymbol{q}}_e$ each for itself to zero. This results in the following set:

- Total system equation

$$\sum_{i=1}^{n} \int_{B_i} \left\{ \left[\left(\frac{\partial \boldsymbol{v}_{oi}}{\partial \dot{\boldsymbol{q}}_r} \right) + (\tilde{\boldsymbol{x}}_i + \tilde{\boldsymbol{s}}_i)^T \left(\frac{\partial \boldsymbol{\omega}_{oi}}{\partial \dot{\boldsymbol{q}}_r} \right) \right]^T d\dot{\boldsymbol{p}}_{Ei} + \left(\frac{\partial \boldsymbol{\omega}_{oi}}{\partial \dot{\boldsymbol{q}}_r} \right)^T d\dot{\boldsymbol{L}}_i \right\} = 0 \tag{2.10}$$

- Elastic link equation

$$\sum_{i=1}^{n} \int_{B_i} \left\{ \left[\left(\frac{\partial \boldsymbol{v}_{oi}}{\partial \dot{\boldsymbol{q}}_e} \right) + (\tilde{\boldsymbol{x}}_i + \tilde{\boldsymbol{s}}_i)^T \left(\frac{\partial \boldsymbol{\omega}_{oi}}{\partial \dot{\boldsymbol{q}}_e} \right) + \left(\frac{\partial \dot{\boldsymbol{s}}_i}{\partial \dot{\boldsymbol{q}}_e} \right) \right]^T d\dot{\boldsymbol{p}}_{Ei} + \right.$$

$$\left. \left[\left(\frac{\partial \boldsymbol{\omega}_{oi}}{\partial \dot{\boldsymbol{q}}_e} \right) + (\boldsymbol{E} + \tilde{\boldsymbol{\varphi}}_i)^T \left(\frac{\partial \boldsymbol{\varphi}_i}{\partial \dot{\boldsymbol{q}}_e} \right) \right]^T d\dot{\boldsymbol{L}}_i \right\} = 0 \tag{2.11}$$

- Abbreviations:

$$d\dot{\boldsymbol{p}}_{Ei} = \left[(\dot{\boldsymbol{v}}_{Ei} + \tilde{\boldsymbol{\omega}}_{oi}\,\boldsymbol{v}_{Ei}) - (\boldsymbol{f}_i + \frac{1}{\rho_i} div\,\boldsymbol{S}_i)\right] dm_i,$$

$$d\dot{\boldsymbol{L}}_i = d\boldsymbol{I}_i\dot{\boldsymbol{\omega}}_i + \tilde{\boldsymbol{\omega}}_i\, d\boldsymbol{I}_i\boldsymbol{\omega}_i - d\boldsymbol{T}_{ai} \tag{2.12}$$

- Dimensions:

$$\boldsymbol{q}_r \in \mathbb{R}^{f_r}\,,\ \boldsymbol{q}_e \in \mathbb{R}^{f_e}\,,$$

$$(\boldsymbol{v}_{oi}\,,\ \boldsymbol{\omega}_{oi}\,,\ \boldsymbol{x}_i\,,\ \boldsymbol{s}_i\,,\ \boldsymbol{\varphi}_i\,,\ \boldsymbol{p}_{Ei}\,,\ \boldsymbol{L}_{Ei}\,,\ \boldsymbol{F}_{ai}\,,\ \boldsymbol{T}_{ai}) \in \mathbb{R}^3$$

$$\left(\frac{\partial \boldsymbol{v}_{oi}}{\partial \dot{\boldsymbol{q}}_r}\right) \in \mathbb{R}^{3,f_r}\,, \qquad \left(\frac{\partial \dot{\boldsymbol{s}}_i}{\partial \dot{\boldsymbol{q}}_e}\right) \in \mathbb{R}^{3,f_e} \text{ etc.} \tag{2.13}$$

The evaluation of these equations is performed numerically, where in a first step it is advisable to collect all integrals containing terms with spatial expressions in certain integral matrices which must be computed before such a numerical integration. One of the typical expression leading to these spatial integral matrices is

$$\sum_{i=1}^{n}\int_{B_i}\left[\left(\frac{\partial \boldsymbol{v}_{oi}}{\partial \dot{\boldsymbol{q}}_e}\right) + (\tilde{\boldsymbol{x}}_i + \tilde{\boldsymbol{s}}_i)^T\left(\frac{\partial \boldsymbol{\omega}_{oi}}{\partial \dot{\boldsymbol{q}}_e}\right)\right]^T\left[\frac{1}{\rho_i} div \boldsymbol{S}_i\right] dm_i = -\left(\frac{\partial V_i}{\partial \boldsymbol{q}_e}\right)^T + \sum_k \boldsymbol{F}_{ik}(\boldsymbol{q}). \tag{2.14}$$

The elastic Potential V_i follows from

$$\boldsymbol{S} = \boldsymbol{H}\,\epsilon, \tag{2.15}$$

$$V_i = \frac{1}{2}\int_{B_i}\left(\epsilon^T \boldsymbol{H}\,\epsilon\right) dV. \tag{2.16}$$

For a linearization of eq.(2.10) and eq.(2.11) we must develop eq.(2.12) for $d\boldsymbol{p}_{Ei}$ and $d\boldsymbol{L}_i$ as well as the bracket expressions of each eq.(2.10), eq.(2.11) up to first order terms in the elastic coordinates $\boldsymbol{q}_e$ for not loosing an elastic contribution to the gross motion of the robot.

The combination of zero order forces or torques in eq.(2.12) with first order elastic terms in the brackets of eq.(2.10), eq.(2.11) is sometimes called "stiffening effect" due to their physical meaning in special applications. Developing the Jacobians in eq.(2.10), eq.(2.11) up to first order terms means developing the velocity constraints up to second order terms, which might be quadratic terms either in $(\boldsymbol{q}_e, \dot{\boldsymbol{q}}_e)$ or in $(\boldsymbol{q}_e, \boldsymbol{q}_e)$ alone. Looking at eq.(2.1), for example, elastic coordinates enter the expressions of $(\boldsymbol{v}_{oi}, \boldsymbol{\omega}_{oi})$ due to the deformations of all bodies down to the frame I, and they enter through $(\boldsymbol{s}_i, \dot{\boldsymbol{s}}_i, \boldsymbol{\varphi}_i)$. Usually, the velocities $(\boldsymbol{v}_{Ei}, \boldsymbol{\omega}_{Ei})$ depend only linearly on $(\dot{\boldsymbol{q}} = \dot{\boldsymbol{q}}_r, \dot{\boldsymbol{q}}_e)$. Therefore, we must focus our attention to developing the terms in eq.(2.10), eq.(2.11) in a Taylor series with respect to $\boldsymbol{q}_e$. A typical example would be the radius vector $\boldsymbol{r}_{oi}$ from the inertial frame I to the frame B_{oi}, Fig. 2.1. The $\boldsymbol{r}_{oi}$ contains all deformations of the bodies between I and B_{oi}. Up to second order we get by a Taylor expansion:

$$\boldsymbol{r}_{oi}(\boldsymbol{q}_r, \boldsymbol{q}_e) = \boldsymbol{r}_{ooi}(\boldsymbol{q}_r) + \left(\frac{\partial \boldsymbol{r}_{oi}}{\partial \boldsymbol{q}_e}\right)_{\boldsymbol{q}_e=0} \boldsymbol{q}_e + \frac{1}{2}\boldsymbol{q}_e^T \left(\frac{\partial}{\partial \boldsymbol{q}_e}\left(\frac{\partial \boldsymbol{r}_{oi}}{\partial \boldsymbol{q}_e}\right)_{\boldsymbol{q}_e=0}\right) \boldsymbol{q}_e + \cdots \tag{2.17}$$

Similar expansions must be performed for all magnitudes containing $\boldsymbol{q}_e$ or $\dot{\boldsymbol{q}}_e$. Doing so we come out with a set of equations of motion which describe the reference motion as performed by the robot with rigid components only and a second set describing the elastic deviations

- Reference Motion:

$$\boldsymbol{M}_o(\boldsymbol{q}_o, t)\ddot{\boldsymbol{q}}_o + \boldsymbol{f}_o(\boldsymbol{q}_o, \dot{\boldsymbol{q}}_o, t) = \boldsymbol{B}_o \boldsymbol{T}_o(t) + \boldsymbol{J}_T^T \boldsymbol{F}_{Go} + \boldsymbol{J}_R^T \boldsymbol{T}_{Go} \tag{2.18}$$

- Linearization:

$$\boldsymbol{q} = \boldsymbol{q}_o + \boldsymbol{y}\,, \boldsymbol{T} = \boldsymbol{T}_o + \Delta \boldsymbol{T}\,, \boldsymbol{F}_G = \boldsymbol{F}_{Go} + \boldsymbol{f}_G\,, \boldsymbol{T}_G = \boldsymbol{T}_{Go} + \boldsymbol{t}_G \tag{2.19}$$

- Deviation Dynamics:

$$\begin{aligned} &\boldsymbol{M}_o(\boldsymbol{q}_o)\ddot{\boldsymbol{y}} + \boldsymbol{P}_o(\boldsymbol{q}_o, \dot{\boldsymbol{q}}_o)\dot{\boldsymbol{y}} + \boldsymbol{Q}_o(\boldsymbol{q}_o, \dot{\boldsymbol{q}}_o, \ddot{\boldsymbol{q}}_o)\boldsymbol{y} = \\ &= \boldsymbol{h}_o(\boldsymbol{q}_o, \dot{\boldsymbol{q}}_o, \ddot{\boldsymbol{q}}_o) + \boldsymbol{B}_o \Delta \boldsymbol{T} + \boldsymbol{J}_T^T(\boldsymbol{q}_o)\boldsymbol{f}_G + \boldsymbol{J}_R^T(\boldsymbol{q}_o)\boldsymbol{t}_G \end{aligned} \tag{2.20}$$

- Dimensions:

$$\begin{aligned} \boldsymbol{y} &= [\boldsymbol{y}_{MI}^T, \boldsymbol{y}_{ME}^T, \boldsymbol{y}_e]^T \in \mathbb{R}^f\,, f = f_r + f_e\,, (\boldsymbol{y}_e = \boldsymbol{q}_e) \\ \boldsymbol{q}_o &= [\boldsymbol{q}_{o,MI}^T, \boldsymbol{q}_{o,ME}^T, 0]^T \in \mathbb{R}^f \end{aligned} \tag{2.21}$$

The reference motion includes a reference trajectory $\boldsymbol{q}_o$ and reference forces and torques $(\boldsymbol{F}_{Go}, \boldsymbol{T}_{Go})$ at the gripper [22]. $\boldsymbol{T}_o(t)$ are the joint torques. The linearization includes the above described one with respect to elasticity and one with regard to forces and torques. For deviation dynamics follows a linear time-variant set (eq.(2.20)), where all matrices and vectors depend on the reference dynamics. The magnitudes $(\boldsymbol{q}_{o,MI}, \boldsymbol{q}_{o,ME})$ are internal and external joint coordinates, $(\boldsymbol{y}_{MI}, \boldsymbol{y}_{ME})$ deviations from their reference values due to joint elasticity. For an ideal robot, eq.(2.18) would be performed perfectly, and the robot would follow some given trajectory with prescribed forces (Fig. 2.2). For a real robot the deviation dynamics must additionally be counterbalanced to achieve a good reference performance.

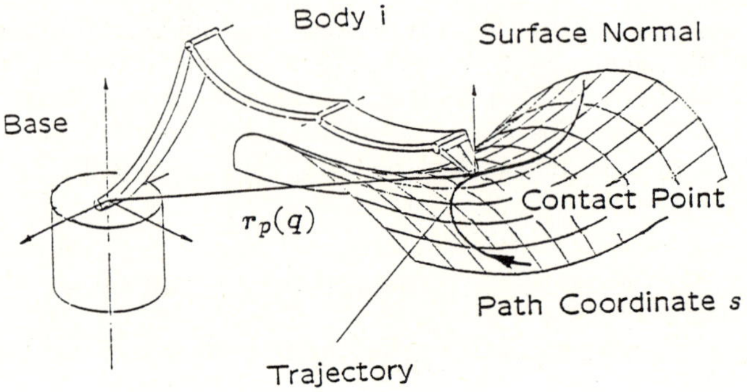

Fig. 2.2: Closed Loop Motion of an Elastic Robot

2.2 Closed Loop Conditions

The equations of motion (2.18), (2.20) follow the main concept of an elastic manipulator following a prescribed trajectory with as well prescribed forces and torques at the gripper, the motion of which is composed of a nonlinear approximation of the system with rigid links and elastic joints superimposed by a linear deviational motion due to elasticities of links and joints. Therefore, it is sufficient to regard an additional closing loop condition only in the equations of the deviation dynamics. For a robot with rigid links the transformation from some body-fixed frame B_{oi} (Fig. 2.1) to an inertial frame I is performed by a matrix $\boldsymbol{R}_i$, which can be composed recursively by the elementary joint matrices $\boldsymbol{T}_{ri}$ [22]. Adding elasticity according to Fig. 2.2 makes it necessary to add the rotational deformation matrix $\tilde{\boldsymbol{\varphi}}_{oi}^T$ and to modify the lentgh $\bar{l}_{oi}$ of a link (Fig. 2.2). With the vectors of Fig. 2.1 we obtain for the chain of Fig. 2.2 transformed to the inertial base:

$$\begin{aligned} \boldsymbol{r}_P(\boldsymbol{q}_o, \boldsymbol{y}) &= \sum_{i=1}^{f_j} \left[\prod_{k=1}^{i} (\boldsymbol{T}_{rk} + \tilde{\boldsymbol{\varphi}}_{ok}^T) \right] (\boldsymbol{x}_i + \boldsymbol{s}_i)_{li}, \text{where } \boldsymbol{T}_{rk} = \boldsymbol{T}_{rk}(\boldsymbol{q}_o, \boldsymbol{y}_{ME}), \\ \tilde{\boldsymbol{\varphi}}_{ok} &= \tilde{\boldsymbol{\varphi}}_{ok}(\boldsymbol{y}_e), \boldsymbol{s}_i = \boldsymbol{s}_i(\boldsymbol{y}_e). \end{aligned} \tag{2.22}$$

Neglecting all deformation terms higher than first order gives:

$$\boldsymbol{r}_P(\boldsymbol{q}_o + \boldsymbol{y}) = \boldsymbol{r}_{Po}(\boldsymbol{q}_o) + \left(\frac{\partial \boldsymbol{r}}{\partial \boldsymbol{q}} \right)_o \boldsymbol{y}, \tag{2.23}$$

where $(\partial \boldsymbol{r} / \partial \boldsymbol{q})_o$ can be interpreted as the Jacobian matrix of translation with respect to the complete coordinate vector $\boldsymbol{y}$ defined in eq.(2.21). The geometrical constraint in the direction of the normal vector $\boldsymbol{n}$ is given by

$$\boldsymbol{n}^T (\boldsymbol{J}_T \boldsymbol{y}) = 0, \tag{2.24}$$

a condition, which closes the loop and which at the same time reduces the force deviations $(\boldsymbol{f}_G, \boldsymbol{t}_G)$ in eq.(2.20). To combine eq.(2.20) with eq.(2.24) we differentiate eq.(2.24) twice to give

$$\boldsymbol{w}^T\ddot{\boldsymbol{y}} + 2\dot{\boldsymbol{w}}^T\dot{\boldsymbol{y}} + \ddot{\boldsymbol{w}}^T\boldsymbol{y} = 0, \text{ with } \boldsymbol{w}^T = \boldsymbol{n}^T\boldsymbol{J}_T,\ \dot{\boldsymbol{w}}^T = \boldsymbol{n}^T\dot{\boldsymbol{J}}_T,\ \ddot{\boldsymbol{w}}^T = \boldsymbol{n}^T\ddot{\boldsymbol{J}}_T. \tag{2.25}$$

Equations (2.20) and eq.(2.25) may now be written in the standard forms

$$\boldsymbol{M}(\boldsymbol{q},t)\ddot{\boldsymbol{y}} - \boldsymbol{h}^*(\boldsymbol{y},\dot{\boldsymbol{y}},t) = 0, \quad \boldsymbol{w}^T(\boldsymbol{q},t)\ddot{\boldsymbol{y}} - \zeta(\boldsymbol{y},\dot{\boldsymbol{y}},t) = 0. \tag{2.26}$$

Applying a Lagrangian approach with a constraint force magnitude λ gives

$$\boldsymbol{M}\ddot{\boldsymbol{y}} + \boldsymbol{w}\lambda = \boldsymbol{h}^*, \qquad \boldsymbol{w}^T\ddot{\boldsymbol{y}} = \zeta \tag{2.27}$$

which results in

$$\ddot{\boldsymbol{y}} = \boldsymbol{M}^{-1}(\boldsymbol{h}^* - \boldsymbol{w}\lambda), \qquad \lambda = (\boldsymbol{w}^T\boldsymbol{M}^{-1}\boldsymbol{w})^{-1}(\boldsymbol{w}^T\boldsymbol{M}^{-1}\boldsymbol{h}^* - \zeta). \tag{2.28}$$

The equations (2.27) give explicitely the unknown deviation accelerations and the constraint force due to the closing condition. To realize the reference eqs. (2.18) it is therefore sufficient to nullify λ by some control concept, which means that the deviations $(\boldsymbol{f}_G, \boldsymbol{t}_G)$ need not to be considered in the vector $\boldsymbol{h}^*$ of eq.(2.26). Therefore,

$$\boldsymbol{h}^* = \boldsymbol{h}_o + \boldsymbol{B}_o\Delta\boldsymbol{T}. \tag{2.29}$$

From this and from the equations (2.20), (2.25), (2.26), (2.27) we get the final set of an elastic robot with a closed loop:

$$\boldsymbol{M}\ddot{\boldsymbol{y}} + \boldsymbol{P}\dot{\boldsymbol{y}} + \boldsymbol{Q}\boldsymbol{y} = \boldsymbol{h} + \boldsymbol{B}\Delta\boldsymbol{T} \tag{2.30}$$

$$\lambda = m_c[\boldsymbol{w}^T\boldsymbol{M}^{-1}(\boldsymbol{h}_o + \boldsymbol{B}_o\Delta\boldsymbol{T} - (\boldsymbol{w}^T\boldsymbol{M}^{-1}\boldsymbol{P}_o - 2\dot{\boldsymbol{q}}^T)\dot{\boldsymbol{y}} - (\boldsymbol{w}^T\boldsymbol{M}^{-1}\boldsymbol{Q}_o - \ddot{\boldsymbol{w}}^T)\boldsymbol{y}] \tag{2.31}$$

$$\begin{aligned} \boldsymbol{P} &= \hat{\boldsymbol{W}}\boldsymbol{P}_o + \boldsymbol{w}m_c(2\dot{\boldsymbol{w}}^T),\ \boldsymbol{Q} = \hat{\boldsymbol{W}}\boldsymbol{Q}_o + \boldsymbol{w}m_c(\ddot{\boldsymbol{w}}^T),\ \boldsymbol{h} = \hat{\boldsymbol{W}}\boldsymbol{h}_o \\ \boldsymbol{B} &= \hat{\boldsymbol{W}}\boldsymbol{B}_o,\ m_c = (\boldsymbol{w}^T\boldsymbol{M}^{-1}\boldsymbol{w})^{-1},\ \hat{\boldsymbol{W}} = \boldsymbol{E}_f - \boldsymbol{w}(m_c\boldsymbol{w}^T\boldsymbol{M}^{-1}) \end{aligned} \tag{2.32}$$

3 Examples

3.1 Bar-Like Structures

Analyzing Fig. 2.1 in more detail and considering classical results of torsion and bending of Euler-Bernoulli-bars we get the following expressions up to second order terms with regard to the elastic deformations:

$$\boldsymbol{s} = \boldsymbol{s}_1 + \boldsymbol{s}_2$$

$$\boldsymbol{s}_1 = \begin{pmatrix} u \\ v \\ w \end{pmatrix}; \quad \boldsymbol{s}_2 = \begin{pmatrix} -\frac{1}{2}\int_o^x \left(v'^2 + w'^2\right) d\xi \\ \int_o^x\int_o^\xi w''\alpha d\eta d\xi + \int_o^x u'v' d\xi \\ \int_o^x\int_o^\xi v''\alpha d\eta d\xi + \int_o^x u'w' d\xi \end{pmatrix} \tag{3.1}$$

$$\boldsymbol{T} = \boldsymbol{E} + \boldsymbol{T}_1 + \boldsymbol{T}_2$$

$$\boldsymbol{T}_1 = \tilde{\boldsymbol{\varphi}}^T = \begin{pmatrix} 0 & -v' & -w' \\ v' & 0 & -\alpha \\ w' & \alpha & o \end{pmatrix}$$

$$\boldsymbol{T}_2 = \begin{pmatrix} -\frac{1}{2}\int_o^x (v'^2 + w'^2) & -\int_o^x \alpha' w' d\xi & -\int_o^x \alpha' v' d\xi \\ -\int_o^x \alpha w'' d\xi & -\frac{1}{2}\int_o^x (v'^2 + \alpha'^2) & -\int_o^x v'w'' d\xi \\ +\int_o^x \alpha v'' d\xi & +\int_o^x v''w' d\xi & -\frac{1}{2}\int_o^x (\alpha'^2 + w'^2) \end{pmatrix} \tag{3.2}$$

The elastic potential V in eq.(2.14) can be expressed by

$$V = \int_o^L (u', v'', w'', \alpha') \begin{pmatrix} EA & 0 & 0 & 0 \\ 0 & E\overline{I}_z & 0 & 0 \\ 0 & 0 & E\overline{I}_y & 0 \\ 0 & 0 & 0 & G\overline{I}_t \end{pmatrix} \begin{pmatrix} u' \\ v'' \\ w'' \\ \alpha' \end{pmatrix} dx, \tag{3.3}$$

where $(EA, E\overline{I}_z, E\overline{I}_y, G\overline{I}_t)$ characterize longitudinal, bending and torsional rigidity. A well-known effect is the stiffening of a rotating bar due to centrifugal forces, which can only be evaluated by taking into account second order terms.

3.2 Elastic Manipulator with Force Control

We consider an elastic manipulator following a prescribed path with a prescribed force [22], [6]. The path is controlled in the tangent plane of each trajectory point lying on a given surface (Fig. 2.2), the force is controlled (including dry friction forces) in a plane vertical to that tangent surface plane, but tangent to the velocity vector in the point under consideration. The multistage concept is described in [6], [22]. Fig. 3.1 gives a typical force-controlled trajectory on a plane table and Fig. 3.2 a comparison of measured and simulated beam stresses in some links. Agreement is excellent. More detailed results may be found in [6], [22].

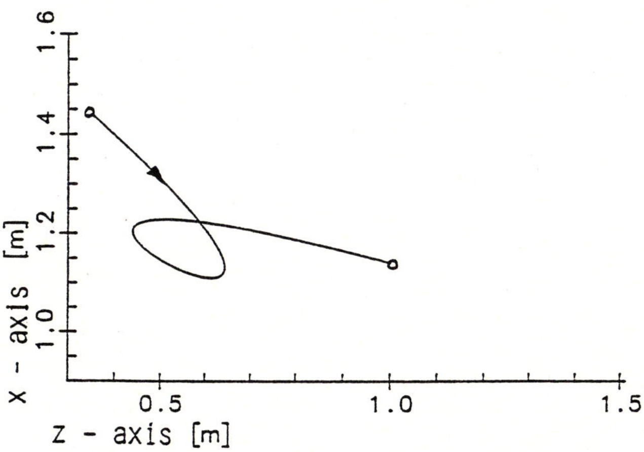

Fig. 3.1: Force-Controlled Trajectory on a Plane

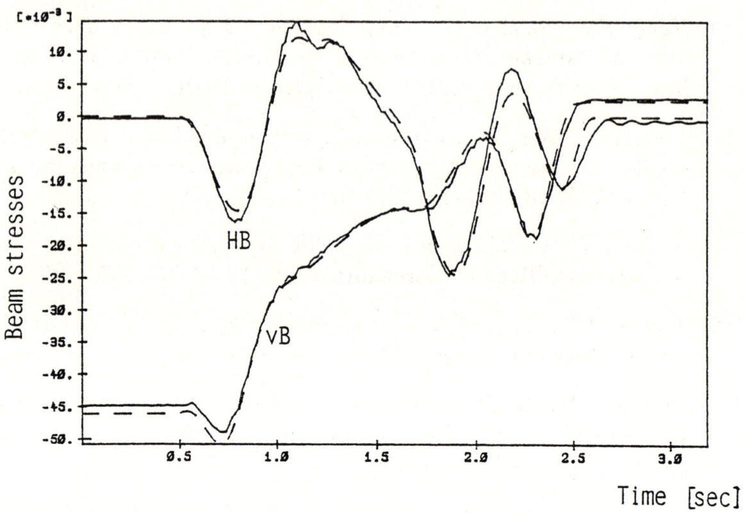

Fig. 3.2: Vertical (VB) and Horizontal (HB) Bending of the First Link (- - - Theory, — Measurements)

References

[1] d'Alembert, Jean le Rond: **Traité de Dynamique.** Chez David, Libraire, Paris 1758

[2] Brandl,H., Johanni, R..; Otter, M.: **A Very Efficient Algorithm for the Simulation of Robots and Similar Multibody Systems without Inversion of the Mass Matrix.** IFAC/IFIP/IMACS Symp. on Theory of Robots, Vienna/Austria, Dec. 1986

[3] Bremer, H.: **Dynamik und Regelung mechanischer Systeme**, Teubner, Stuttgart, 1988.

[4] Bremer, H.: **On the Dynamics of Flexible Manipulators.** Proc. 2nd IEEE Conf. Robotics and Automation, Rayleigh/USA, 1556 - 1560, 1987

[5] Bremer, H.: **Dynamics of Multibody Systems with Elastic Components.**(in German), ZAMM 65, 613 - 621, 1985

[6] Bremer, H., Pfeiffer, F.: **Control of Elastic Robots.** Proc. IUTAM Symp.on Dynamic Problems of Rigid-Elastic Systems and Structures, Moscow, May 23 - 27, to appear

[7] Fleischer, G.C.; Likins, P.W.: **Attitude dynamics simulation subroutines for systems of hinge-connected rigid bodies.** JPL Tech. Report 32-1592, Pasadena, 1974

[8] Gebler, B.: **Modellbildung, Regelung und Steuerung für elastische Industrieroboter.** VDI-Fortschr.-Ber, Reihe 11, Nr. 98, Düsseldorf 1987

[9] Haug, E.J.; Shic, C.W.; Sung-Soo Kim: **Dynamics of Flexible Machines: A Variational Approach.** IUTAM/IFToMM Symp. on Dynamics of Multibody Systems, Udine 1985, Springer Berlin Heidelberg 1986

[10] Hooker, W.W.: **Equations of motion for interconnected rigid and elastic bodies: a derivation independent of angular momentum.** Celestial Mech. 2 (1975), 337 - 359

[11] Hooker, W.W.; Margulies, G.: **The dynamics attitude equations for n-body satellite.** J. Astronautical Sci. 12 (1965), 123 - 128

[12] Huston, R.L.; Pasarello, C.E.; Harlow, M.W..: **Dynamics of Multi-Rigid-Body Systems.** J. Appl. Mech. 45 (1978), 889 - 894

[13] Jourdain, P.E.B.: **Note on an Analogue of Gauss' Principle of Least Constraint.**, Quarterly J. of Pure and Appl. Math., Vol. XL, London, Longmans, Green and Co., 1909

[14] Kane, T.R.; Levinson, D.A.: **Dynamics: Theory and Application.** New York, McGraw Hill, 1985

[15] Kane, T.R.: **Dynamics of Nonholonomic Systems.** J. Appl. Mech. 28 (1961), 574 - 578

[16] Kleemann, U.: **Regelung elastischer Roboter**, VDI Fortschr.-Ber., Reihe 8, Nr. 191, Düsseldorf 1989.

[17] Lachenmayr, G.: **Schwingungen in Plantetengetrieben mit elastischen Hohlrädern.** VDI Fortschr.-Ber., Reihe 11, Nr. 108, Düsseldorf 1988

[18] Lagrange, J.-L.: **Mécanique Analytique.** Veuve Desaint, Paris 1788, Reprint Editions Jaques Gabay, Sceaux 1989

[19] Lilov, L.K.: **Dynamics of Elastic Multibody Systems Involving Closed Loops.** IUTAM/IFToMM Symp. on Dynamics of Multibody Systems, Udine 1985, Springer Berlin Heidelberg 1989

[20] Pfeiffer, F.; Gebler,B.: **A Multistage Approach to the Dynamics and Control of Elastic Robots.** Proc. IEEE Int. Conf. on Robotics and Automation, Philadelphia/USA, 2 - 8, 1988

[21] Pfeiffer, F.; Kleemann, U. **Elasticity and Vibration Control for Manipulators.** Proc. IEEE Int. Conf. on Control and Applications, Jerusalem, 1989

[22] Pfeiffer, F.; Richter, K.; Wapenhans, H.: **Elastic Robot Trajectory Planning with Force Control.** Proc. of the IFIP International Symp. on Theory of Robots, Rome, Italy 1990

[23] Schiehlen, W.O.: **Dynamics of Complex Multibody Systems.** SM Archives 9 (1984), 159 - 195

[24] Schwertassek, R.; Roberson, R.E.: **A state-space dynamical representation for multibody mechanical systems.** Acta Mech. 50 1983, 141 - 161 **Part II: Systems with closed loops.** Acta Mech 51 (1984), 15 - 29

[25] Singh, R.P.; Vandervoort, R.J.: **Dynamics of Flexible Bodies in Tree Topology: A Computer Oriented Approach.** J. Guid. Contr., Vol 8, Nr. 5, Sept. 1985, 584 - 590

[26] Ulbrich, H.: **Dynamik und Regelung von Rotorsystemen.** VDI Fortschr.-Ber., Reihe 11, Nr. 86, D"usseldorf 1986

[27] Wittenburg, J.: **The dynamics of systems of coupled rigid bodies. A new general formalism with applications.** Stereodynamics Centro Internazionale Matematica Estivo, I Ciclo 1971, (Bressanone, 2-12 Jun 1971), Edizioni Cremonese, Roma, 1972

[28] Wittenburg, J.: **Dynamics of Systems of Rigid Bodies.** Teubner, Stuttgart 1977

Inverse Problems and Global Optimization of the Oscillatory Systems

V. M. RYABOY

Mechanical Engineering Research Institute,
USSR Academy of Sciences,
4, Griboedova, 101830, Moscow, USSR

Summary

Most of the known methods of structural optimization involve local optimization, i.e., the determination of optimal parameters for the system with specified configuration, or local modification of this configuration [1,2]. In this paper, an approach is presented for global optimization of linear mechanical oscillatory systems using the inverse problem solution for the system with given frequency response. The optimization is conducted in the function space of frequency responses rather than in the system parameters domain. The connection between the two domains is provided by the solution of the inverse problem.

With this approach, the limiting performance of the linear multi-degrees-of-freedom vibration isolation systems is investigated and the procedure for the determination of optimal configuration and parameters is developed.

Problem Statement

Let us consider the mechanical system of any number of material points with masses m_1, m_2, ... , m_n, which are elastically connected with each other and with a rigid foundation. The potential energy of the system is assumed to be a general positive definite quadratic form of the displacements of all the material points. If the system is excited in the point m_1 by a harmonic force with frequency ω and amplitude vector $\{f_1\}$, then the harmonic force is acting on the foundation with the same frequency ω and amplitude vector $\{f_0\}$. The amplitude vectors are related through the linear equation $\{f_0\}=[T(\omega)]\{f_1\}$, where $[T(\omega)]$ is a (3×3) transmissibility matrix which characterizes the vibration isolation properties. The aim of the vibration isolation is to reduce a force acting on the foundation. Hence the vibration isolation system is required to be

N. V. Banichuk · D. M. Klimov · W. Schiehlen (Eds.)
Dynamical Problems of Rigid-Elastic
Systems and Structures
IUTAM Symposium Moscow, USSR, May 23-27, 1990

such that the norm of this matrix is less than the prescribed positive function ε, i.e., $\|T(\omega)\| \leq \varepsilon(\omega)$, in some frequency domain Ω. An equivalent problem is to reduce the displacement of the point m_1 when the foundation is excited kinematically.

Intuitively, the vibration isolation in such a system is due to inertia, while the inertia, in its turn, depends on mass. This intuitive argument and numerous examples suggest that the total mass of a system increases infinitely as the required vibration isolation efficiency increases, i.e., $\varepsilon \to 0$. Our intention is to prove and estimate this feature for the whole class of linear passive systems. Such estimates are of great value for the theory and practice of vibration isolation because they describe the limiting performance of passive systems, and show, whether the use of more complicated and expensive active systems is justified or not.

A rigorous formulation of the problem is presented here under assumption that the system is uni-directional, i.e. each material point can move just in one direction which coincides with a fixed line of action of the excitation force. This does not result in loss of generality [5], but simplifies the presentation. The equation of stationary harmonic motion is

$$[C]\{u\}-\omega^2[M]\{u\}=\{f\}, \tag{1}$$

where [C] is a general positive definite stiffness matrix, [M] is a diagonal mass matrix ($[M] = \mathrm{diag}[m_1,m_2,...,m_n]$), {u} is n-dimensional column vector of displacement amplitudes; $\{f\}=\{f_1,0,...,0\}^T$ is the force amplitude vector. According to D'Alembert's principle and Eq.(1), the amplitude f_0 of the force acting on the foundation is

$$f_0=\{e\}[C]\{u\}, \tag{2}$$

where $\{e\}=\{1,1,...,1\}$. The transmissibility is now a scalar,

$$T(\omega)=f_0/f_1. \tag{3}$$

If the foundation is excited kinematically with a given amplitude u_0 and frequency ω (and there is no force excitation) then $T(\omega)=u_1/u_0$.

Hence the vibration isolation requirement is

$$|T(\omega)| \leq \varepsilon(\omega), \quad \omega \in \Omega, \tag{4}$$

where ε is prescribed function, and Ω is prescribed frequency domain which is called vibration isolation domain. For the problem to be meaningful, some requirement must be added which ensures the integrity of the whole system. In this paper, we require that transmissibility $T(\omega)$ be finite in some low-frequency range,

$$|T(\omega)| < \infty, \qquad 0 \leq \omega < \omega_q, \tag{5a}$$

or, equivalently, the lowest eigenfrequency ω_1 be greater than or equal to prescribed positive ω_q,

$$\omega_1 \geq \omega_q \tag{5b}$$

The constraint (5) is, in fact, the "weak" low-frequency stabilization requirement. Alternative integrity requirements were suggested in [3,5]. The total mass of a system is the sum of all the elements of [M] matrix,

$$m=\{e\}[M]\{e\}^T. \tag{6}$$

Our intention is to estimate the minimum possible value of the total mass m for the given $\varepsilon(\omega)$, Ω, ω_q and mass m_1 which simulates the isolated object. More exactly, the problem is to estimate inf(m) for all the matrices [M] and [C] of any finite dimensions with the aforementioned properties, provided Eqs. (1) to (6) are valid for the fixed m_1, $\varepsilon(\omega)$, Ω and ω_q. It is convenient to present the result in the form of an inequality which relates the total mass with other parameters involved in the formulation. Such a general inequality would describe the limiting performance of all the linear conservative vibration isolation systems.

Solution of inverse and optimal problems

The full description of the system is given by the matrices [M] and [C], while the global properties of a system with the "input" and "output" in the point m_1 and/or on the foundation can be specified by fewer parameters. These parameters are introduced via a modal presentation of the system. The dynamic matrices [M] and [C] are diagonalized simultaneously:

$$[\Phi]^T[M][\Phi]=[E], \quad [\Phi]^T[C][\Phi]=[\Lambda]. \tag{7}$$

The columns of [Φ] matrix are the eigenvectors, [E] is the identity matrix, and $[\Lambda]=\mathrm{diag}[\omega_1^2,\omega_2^2,\ldots,\omega_n^2]$, where ω_k are the eigenfrequencies. The dynamic matrices are readily calculated using the inverse transpose of [Φ] by

$$[M]=[\Psi][\Psi]^T, \quad [C]=[\Psi][\Lambda][\Psi]^T, \tag{8}$$

where $[\Psi]=([\Phi]^T)^{-1}$. Let us consider the first row of the [Φ] matrix and the sum of rows of the [Ψ] matrix:

$$\{x\}=\{\phi^1\}, \quad \{y\}=\{\psi^1\} + \{\psi^2\} + \ldots +\{\psi^n\}. \tag{9}$$

The components x_k, y_k of the {x} and {y} row vectors along with eigenfrequencies define all the system input-output frequency responses. In particular, the transmissibility is expressed as

$$T(\omega) = \sum_{k=1}^{n} \frac{z_k}{1 - \omega^2/\omega_k^2}, \tag{10}$$

where

$$z_k=x_k y_k \quad (k = 1,\ldots,n), \tag{11}$$

and the following identity holds:

$$\sum_{k=1}^{n} z_k = 1. \tag{12}$$

Moreover, the total system mass m and mass m_1 lumped in the excitation point are defined by these parameters:

$$m = \sum_{k=1}^{n} y_k^2, \quad m_1 = (\sum_{k=1}^{n} x_k^2)^{-1}. \tag{13}$$

The formulas for other global parameters such us bulk stiffness, etc.,are presented in [3-5].

The essential point is that the {x} and {y} parameters form the input data for reconstruction of [M] and [C] matrices. These matrices are readily computed using Eqs. (8), if the [Ψ] matrix and the eigenvalues are known. The $[\Psi]$ matrix is constructed by rows. To do this, one expands $\{y\}$ into the sum of orthogonal vectors $\{\psi^k\}$ according to $(9)_2$, where the first vector $\{\psi^1\}$ is known (it is defined by $\{x\}$: $\{\psi^1\}=\{x\}/|x|^2$) [3,4]. This expansion is the main step of the inverse problem solution. Of course, it is not unique (if $n>2$). This means that the same frequency response functions may belong to a multiplicity of mechanical systems. Hence, the optimization problems for the vast classes of complex oscillatory systems are multi-extremum and, as such, difficult to solve. Nevertheless, such problems can be solved using the new set of parameters or, in other words, in the function space of frequency responses functions defined by these parameters.

The first global problem of interest is to determine the minimum possible mass for the system with specified transmissibility $T(\omega)$ and mass m_1 of isolated object. The solution is readily found via Lagrange's minimization of expression $(13)_1$ for m, with Eqs. $(13)_2$ and (11) as constraints. The minimum mass is

$$m_{min} = m_1 \left(\sum_{k=1}^{n} |z_k| \right)^2. \tag{14}$$

The $\{x\}$ and $\{y\}$ vectors are determined uniquely by

$$y_k^2 = m_1 |z_k| \sum_{j=1}^{n} |z_j|, \quad x_k = z_k / y_k \tag{15}$$

($k=1,2,\ldots,n$; the sign of y_k does not matter).

Now, assume that the eigenfrequencies are fixed and the mass is to be minimized under the vibration isolation requirement (4). This problem reduces to the linear constrained Chebyshev approximation problem, which requires minimization of the quantity in parentheses in Eq.(14) with the inequality constraint (4), (10) and equality constraint (12). It is solved using the well-known procedures of approximation, e.g., Remez's algorithm [6]. The conditions are satisfied which guarantee the existence, uniqueness and convergence of this solution. Some examples are given elsewhere [5].

Limiting performance estimate

The third and most difficult global optimization problem under consideration is the limiting performance problem. It requires minimization of the quantity in the right-hand side of (14) for all admissible sets of z_k and ω_k. "Admissible" means that all the conditions (5b), (4), (10), (12), hold. It is easy to find the upper bound estimates of this minimum. However, it is the lower bound estimate of the minimum mass that is of interest for us. To obtain this estimate, we formulate the dual problem, thereby replacing the minimum by the maximum. Then the calculation of the goal function for any trial element gives a lower bound estimate. Such dual problems are well-known for the linear approximation problem [7], i.e., for fixed eigenfrequencies. Following the main steps outlined in [3,5], we can generalize the relevant dual problem of [7] so that its solution provides the lower bound estimate for all admissible sets of eigenfrequencies. According to this duality theory,

$$\sqrt{m/m_1} \geq \mu,$$

where

$$\mu = \sup_{g,g_1} \left\{ -\int_{\Omega} \varepsilon(\omega)\,|dg(\omega)| + g_1 \right\}, \tag{16}$$

$$\left| \int_{\Omega} \frac{dg(\omega)}{1-\left(\frac{\omega}{\omega'}\right)^2} + g_1 \right| \leq 1, \quad \omega' \in \Omega' = [\omega_q, \infty]\setminus\Omega. \tag{17}$$

Here, g belongs to the set of functions with limited variation in domain Ω, and g_1 is a real number. Introducing transformation of variables $\omega^2 = s$, $\omega'^2 = p$, $\Omega \to S$, $\Omega' \to S'$, $\omega^2 dg(\omega) = h(s)ds$, the problem (16), (17) may be rewritten in terms of the analytic function $H(\zeta)$ in all the complex plane except S,

$$H(\zeta) = \int_S \frac{h(s)ds}{s-\zeta}, \quad \operatorname*{Var}_{S'} H(p) \leq 2, \tag{18}$$

where Var denotes the difference between the maximum and

minimum values. The constraint $(18)_2$ and the goal functional (16) are non-differentiable. To avoid this difficulty, we note that conditions (18) are satisfied if $H(\zeta)$ is taken in the form

$$H(\zeta)=1-\cos\left[\sqrt{\zeta-\omega_q^2}\,F(\zeta)\right],\quad F(\zeta)=\frac{1}{\pi}\int_S \frac{f(t)dt}{t-\zeta},$$

while the goal functional can be replaced by the approximate smooth functional for small $\varepsilon(\omega)$ (see [3,5] for similar calculation). Then the extremal $f(t)$ is determined by the conventional variational calculus methods.

The result may be presented in the form of the inequality

$$\sqrt{\frac{m}{m_1}} \geq \frac{V}{2}\exp\left[\frac{2\omega_q}{\pi V}\int_\Omega \ln\frac{\pi\omega_q}{2\varepsilon(\omega)\sqrt{\omega^2-\omega_q^2}}\,\frac{d\omega}{\omega\sqrt{\omega^2-\omega_q^2}}\right](1+r), \tag{19}$$

where

$$V = 1 - \frac{2\omega_q}{\pi}\int_\Omega \frac{d\omega}{\omega\sqrt{\omega^2-\omega_q^2}},$$

which shows how the minimum possible mass ratio depends on the vibration isolation efficiency required and the frequency domains. Here, r is the remainder term (it can be expressed explicitly) which is small for sufficiently small $\varepsilon(\omega)$, i.e., $r=O(1/|\ln \varepsilon_o|)$, if $\varepsilon=\varepsilon_o\delta_o(\omega)$, δ_o is a fixed function, and $\varepsilon_o\to 0$. It follows from (19) that m/m_1 tends to infinity if $\varepsilon\to 0$. Also the mass ratio increases as the Ω domain is extended. If ω_a is the lower bound of Ω ,and $\omega_q<\omega_a$, then m/m_1 tends to infinity if $\omega_q\to\omega_a$. It is interesting to note that the inequality (19) enables us to assess the possibility of vibration isolation in the under-resonant frequency domain, i.e., for $\Omega\subset[0,\omega_q]$ as well, because it holds for $\Omega=[\omega_a,\omega_q]$, $\omega_a<\omega_q$.

The use of inequality (19) is illustrated in Fig.1. Here, $\Omega=[\omega_a,\omega_b]$ and ε=const. Solid curves correspond to the right-hand side of (19) for various values of ω_b/ω_a and ω_a/ω_q. The points falling below these curves cannot correspond to any linear conservative systems, no matter how sophisticated they are. The dashed lines are conventional dynamic anti-resonant vibration isolators with the tuned mass-spring oscillators. Symbols denote three series of points corresponding to mechanical systems which are synthesized through the procedure outlined in the next section. Here, n varies from 2 (on the abscissa axis) to 8. These results show that the estimate (19) is fairly accurate, since the mechanical systems do exist with performance close to our estimate of the limiting one.

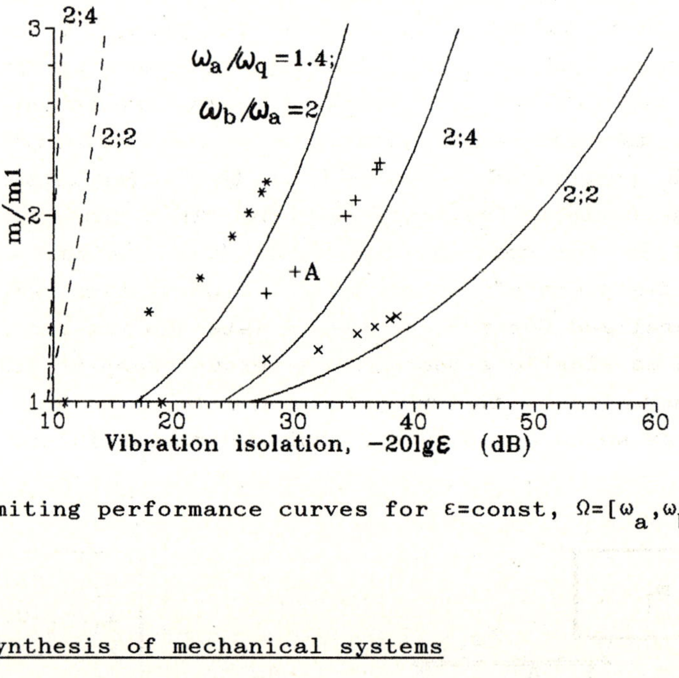

Fig.1. Limiting performance curves for ε=const, $\Omega=[\omega_a,\omega_b]$

Optimal synthesis of mechanical systems

In applications, the problem of the synthesis of system configuration and parameters is of great importance. Its solution can be divided into 5 steps. First, some eigenfrequencies are found based on the dual problem solution. Second, the optimal transmissibility function is found using the approximation algorithm. Then the {x} and {y} parameters are calculated by Eqs.(15). The next step is to calculate [Ψ] with expansion $(9)_2$ and [M], [C] with formulas (8). This step is generally not unique, but it becomes unique if the additional assumption is made that [C] is three-diagonal, i.e., that every material point is connected only with the preceding and subsequent points (and maybe with the foundation). This unique solution is determined using the modified Lancos's algorithm [5]. Finally, the dynamic diagrams composed of the lumped masses, idealized springs and levers are constructed [4,5]. This last step of theoretical mechanics consideration is, in fact, the fist step of the engineering design. There is usually a variety of dynamic diagrams corresponding to given matrices [M] and [C]. Fig 2 is an example of such a diagram.

In practical applications, one is faced with more specific problems. For example, it is necessary to construct a system with increased vibration isolation efficiency in the vicinity of the prescribed frequencies using the least possible number of additional degrees of freedom. In this case an interpolation algorithm is used which gives the transmissibility function $T(\omega)$ with prescribed zeros, while the eigenfrequencies are varied in certain frequency domains. This procedure is implemented in the computer code SYNT [5,8] which is a useful tool for designers of vibration isolation systems. Also, it has been generalized for the case of isolated object and foundation simulated as elastic structures. Numerous examples are given in [5]. In many cases the use of this code gives original configurations which would be difficult to devise intuitively.

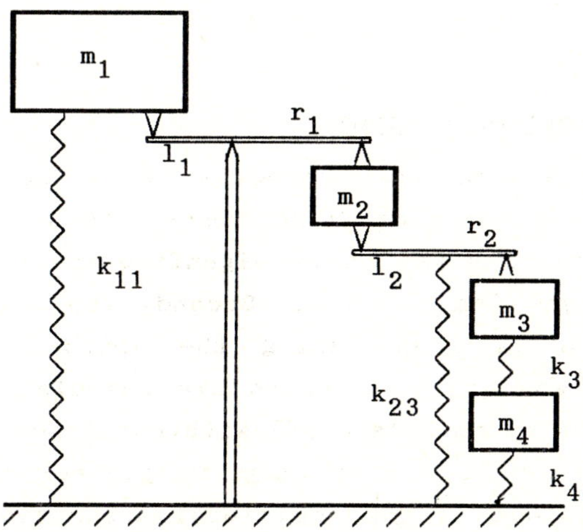

m_1	=1.0
l_1/r_1	=0.301
k_{11}	=0.284
m_2	=0.0086
l_2/r_2	=7.013
k_{23}	=4.829
m_3	=0.180
k_{34}	=0.237
m_4	=0.513
k_{44}	=0.127

Fig.2. One of the dynamic diagrams corresponding to the point A of Fig.1. (ω_b/ω_a=4, ω_a/ω_q=2). ω_a=1, m_1=1

Conclusions

Optimization problems for general classes of oscillatory systems are multi-extremum ones. However, they may be uni-extremum in the function space of frequency response functions. This feature makes it possible to conduct the global optimization. The frequency response functions and the system configuration and parameters are connected through the inverse problem solution.

The global optimum estimate for general linear conservative systems obtained with this approach shows that the performance of all the passive linear elastic-inertial vibration isolation systems is limited principally by the very nature of this class of systems irrespective to their configurations.

The optimization procedure based on this inverse problem approach can provide the designer with concepts of new designs which could not be found intuitively.

References

1. Banichuk, N.V. Problems and methods of optimal structural design. New York: Plenum Press 1983.

2. Haug, E.J.; Choi, K.C.; Komkov V. Design sensitivity analysis of structural systems. Orlando e.a.: Academic Press 1986.

3. Ryaboĭ, V.M.: Limiting capabilities of elastic-inertial vibration-proofing. Mech. Solids 17 (1982) no. 5, 35-42. (Translated from Russian.)

4. Ryaboi, V.M.: Construction of dynamic diagrams and determination of the parameters of vibration-isolating systems with given properties. Sov. Mach. Sci. 1985 no.5 , 37-43. (Translated from Russian.)

5. Genkin, M.D.; Ryaboy, V.M. Elastic-inertial vibration isolation systems. Limiting performance, optimal structures. Moscow: Nauka 1988 (in Russian).

6. Laurent, P.-J. Approximation et optimisation. Paris: Hermann 1972.

7. Gol'stein, E.G. Duality theory in mathematical programming and it's applications. Moscow: Nauka 1971 (in Russian).

8. Genkin, M.D.; Ryaboy, V.M.: Limiting performance estimates and computer-aided synthesis of elastic-inertial vibration isolation systems. 13-th Internat. Congr. on Acoustics. Beograd, 1989. V.3, 341-344.

Simulation and Animation of Multibody Systems

W. Schiehlen

Institute B of Mechanics
University of Stuttgart
Stuttgart, FRG

Summary

For the dynamical analysis of discrete mechanical systems the method of multibody systems is well qualified. The resulting equations of motion are solved by simulation codes and displayed by computer animation. The paper presents an approach for the efficient computation of motions and reactions used for strength estimation and shape optimization including rigid- elastic beam systems.

Introduction

Many rigid-elastic systems and structures like vehicles, spacecrafts or robots, respectively, can be modelled as multibody systems. The complexity of the dynamical equations called for the development of computer-aided formalisms started a quarter of a century ago. The state-of-the-art was presented at IUTAM symposia and other meetings, documented in the corresponding proceedings, see Magnus [1], Haug [2], Bianchi and Schiehlen [3]. The computational aspects have been treated by Kreuzer and Schiehlen [4], Brandl, Johanni and Otter [5], Führer [6] and Valasek [7]. Computer codes available to the public are presented in Ref. [8].

Multibody systems consist of material bodies, constraint elements like joints, bearings and supports, and force elements realized by springs, dampers or servomotors, respectively. A material body is characterized by its rigid body motion and, possibly, by an elastic deformation. Using the concept of the

N.V. Banichuk · D. M. Klimov · W. Schiehlen (Eds.)
Dynamical Problems of Rigid-Elastic
Systems and Structures
IUTAM Symposium Moscow, USSR, May 23-27, 1990

superelement as introduced by Rauh and Schiehlen [9,10], a flexible beam can be interpreted as a series of rigid bodies and elastic force elements, Fig. 1. The parameters of the superelements are found by comparison with the Timoshenko beam [11]. Therefore, the restriction to rigid body systems is justified.

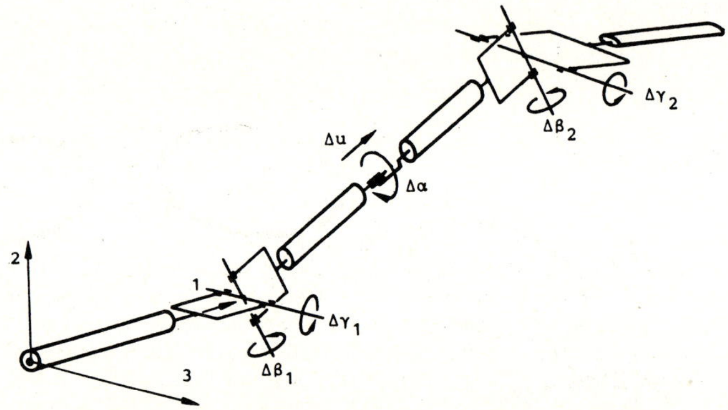

Fig. 1. Rigid-elastic superelement

In particular, at the boundaries of the superelements the reaction forces may be obtained and used for strength estimation and shape optimization, see Ref. [12].

Kinematics and Kinetics of Rigid Body Systems

Rigid-elastic systems consisting of rigid bodies and superelements can always be interpreted as a topological tree iff one superelement in each topological loop shows axial elasticity. The tree structure is equivalent to topological chains which will be analysed in more detail.

A chain system includes rigid bodies characterized by absolute coordinates and joints represented by relative coordinates. For the 3x1-translation vektors r and the 3x3-rotation tensors S of bodies K_1, K_2 and joint J_2 it yields

$$r_{01}(x_2) = r_{01}(x_1) + r_{01P1} + r_{12}(y_2) \ , \tag{1}$$

$$S_{02}(x_2) = S_{01}(x_1)\ S_{01P1}\ S_{12}(y_2) \ , \tag{2}$$

where x_1, x_2 are 6x1- vectors of absolute coordinates and y_2 is the f_{12}x1-vector representing the f_{12} relative coordinates of joint J_2, see Fig. 2.

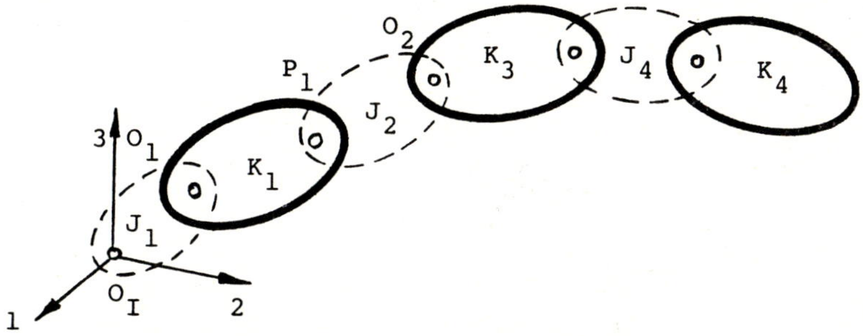

Fig. 2. Kinematics of chain system

By differentiation the following 3x1-vectors for the translational and rotational velocity are obtained,

$$v_{02} = v_{01} - (\widetilde{r_{01P1} + r_{12}})\ \omega_{01} + J_{T2}\ \dot{y}_2 \ , \tag{3}$$

$$\underbrace{\omega_{02}}_{\dot{x}_2} = \underbrace{\qquad\qquad \omega_{01}}_{C_2(y_2)\ \dot{x}_1} + \underbrace{J_{R2}\ \dot{y}_2}_{J_2\ \dot{y}_2} \ , \tag{4, 5}$$

where J_{T2}, J_{R2} are 3xf_{12}- Jacobian matrices, see e.g. Ref. [13], and C_2 is a 6x6- geometry matrix.

For the total system of p bodies, it follows according to (5),

$$\dot{x} = C\ \dot{x} + J\ \dot{y} \tag{6}$$

where the global 6px1-vector x of absolute coordinates and the global fx1-vector y of relative coordinates are connected by the

6px6p-geometry matrix and a block diagonal 6pxf-Jacobian matrix:

$$\mathbf{C} = \begin{bmatrix} 0 & 0 & 0 & \cdots & 0 \\ C_2 & 0 & 0 & \cdots & 0 \\ 0 & C_3 & 0 & \cdots & 0 \\ \vdots & \vdots & \vdots & \ddots & \vdots \\ 0 & 0 & 0 & \cdots & 0 \end{bmatrix}, \quad \mathbf{J} = \begin{bmatrix} J_1 & 0 & 0 & \cdots & 0 \\ 0 & J_2 & 0 & \cdots & 0 \\ 0 & 0 & J_3 & \cdots & 0 \\ \vdots & \vdots & \vdots & \ddots & \vdots \\ 0 & 0 & 0 & \cdots & J_p \end{bmatrix}. \tag{7}$$

From (6) it follows easily

$$\dot{\mathbf{x}} = (\mathbf{E} - \mathbf{C})^{-1} \mathbf{J} \dot{\mathbf{y}} = \bar{\mathbf{J}} \dot{\mathbf{y}}, \tag{8}$$

where the global 6pxf-Jacobian matrix is given by

$$\bar{\mathbf{J}} = \begin{bmatrix} J_1 & 0 & 0 & \cdots & 0 \\ C_2J_1 & J_2 & 0 & \cdots & 0 \\ C_3C_2J_1 & C_3J_2 & J_3 & \cdots & 0 \\ \vdots & \vdots & \vdots & \ddots & \vdots \\ * & * & * & \cdots & J_p \end{bmatrix}. \tag{9}$$

Note that the global 6px6p-Jacobian matrix has triangular form.

In addition the reaction forces acting in a chain system will be treated, Fig. 3.

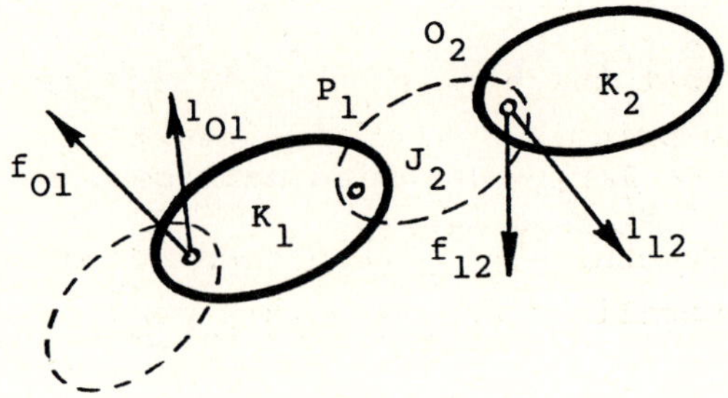

Fig. 3. Reactions in chain system

Body K_1 is affected by joints J_1 and J_2. The resulting reaction forces on body K_1 read as

$$f_1^r = f_{01} - f_{12} \, , \tag{10}$$

$$l_1^r = l_{01} - l_{12} - (r_{01P1} + r_{12}(y_2)) f_{12} \, , \tag{11}$$

$$\bar{q}_1^r = \bar{q}_1 - C_2^T(y_2) \, \bar{q}_2 \, , \tag{12}$$

where the 3x1-vectors f, l represent forces and torques, respectively, while $\bar{q}$ is a 6x1-vector summarizing the forces and torques. Further, the reactions in joint J_1 can be represented by the q_{01}x1-generalized reaction vector g_1 and a corresponding 6pxq_{01}-distribution matrix Q_1 and in joint J_2 similar quantities appear. It yields $f_{12} + q_{12} = 6$ for each joint.

For the total system of p bodies it remains from (12)

$$\bar{q}^r = (E - C)^T Q \, g = \bar{Q} \, g \, , \tag{13}$$

where the global qx1-vector g of generalized reaction forces appear, the block diagonal matrix $Q = \text{diag}\{ Q_i \}$ is used and the global 6px6p-distribution matrix follows as sparse matrix

$$\bar{Q} = \begin{bmatrix} Q_1 & -C_2Q_2 & 0 & \dots & 0 \\ 0 & Q_2 & -C_3Q_3 & \dots & 0 \\ 0 & 0 & Q_3 & \dots & 0 \\ \vdots & \vdots & \vdots & \ddots & \vdots \\ 0 & 0 & 0 & \dots & Q_p \end{bmatrix} . \tag{14}$$

It turns out that there exist a local and global orthogonality according to the principle of virtual work,

$$\delta W^r = \bar{q}^r \, \delta x = g^T \, \bar{Q}^T \, \bar{J} \, \delta y = 0 \, . \tag{15}$$

Since the fx1-vector y of generalized coordinates and the qx1-vektor g of generalized reactions are completely independent it follows

$$\bar{Q}^T \, \bar{J} = 0 \quad \text{and} \quad Q_i^T \, J_i = 0 \, , \quad i = 1(1)p, \tag{16}$$

by definitions (8) and (13).

Equations of Motion and Equations of Reaction

The Newton-Euler equations written down for each rigid body of the system and summarized in global matrix notation read as

$$\bar{\bar{\mathbf{M}}}\,\ddot{\mathbf{x}} + \bar{\mathbf{k}} = \bar{\mathbf{q}}^e + (\mathbf{E} - \mathbf{C})^T \mathbf{Q}\,\mathbf{g}\ , \tag{17}$$

where $\bar{\bar{\mathbf{M}}}$ is a 6px6p-blockdiagonal inertia matrix, $\bar{\mathbf{k}}$ is a 6px1-vector of gyroscopic forces and $\bar{\mathbf{q}}^e$ represents the 6px1-vector of applied forces and torques. Further, from (6) it follows by differentiation

$$\ddot{\mathbf{x}} = \mathbf{C}\,\ddot{\mathbf{x}} + \mathbf{J}\,\ddot{\mathbf{y}} + \xi\ . \tag{18}$$

Thus, there are available 12p differential algebraical equations for the 12p unknowns summarized in the 6px1-vector $\mathbf{x}$, the fx1-vector $\mathbf{y}$ and the qx1-vector $\mathbf{g}$.

In a first step the absolute accelerations (18) are eliminated in (17). Then, it remains

$$\bar{\bar{\mathbf{M}}}(\mathbf{E}-\mathbf{C})^{-1}(\mathbf{J}\ddot{\mathbf{y}}+\xi)+\bar{\mathbf{k}} = \bar{\mathbf{q}}^e+(\mathbf{E}-\mathbf{C})^T\mathbf{Q}\mathbf{g}\ . \tag{19}$$

Now, the orthogonality condition (16) can be applied to generate the equations of motion

$$\bar{\mathbf{J}}^T\,\bar{\bar{\mathbf{M}}}\,\bar{\mathbf{J}}\,\ddot{\mathbf{y}} + \mathbf{k} = \mathbf{q}\ , \tag{20}$$

and the equation of reaction

$$\bar{\mathbf{Q}}^T\,\bar{\bar{\mathbf{M}}}^{-1}\,\bar{\mathbf{Q}}\,\mathbf{g} + \hat{\mathbf{q}} = \hat{\mathbf{k}}\ , \tag{21}$$

where the fx1-vectors $\mathbf{k}$, $\mathbf{q}$ and the qx1-vectors $\hat{\mathbf{k}}$, $\hat{\mathbf{q}}$ represent the influence of gyroscopic and applied forces. Considering the form of matrices (9),(14) and the block diagonality of the inertia matrix $\bar{\bar{\mathbf{M}}}$ it is obvious that $\mathbf{N} = \bar{\mathbf{Q}}^T\bar{\bar{\mathbf{M}}}^{-1}\bar{\mathbf{Q}}$ is a sparse matrix (computation of all reaction forces) while $\mathbf{M} = \bar{\mathbf{J}}^T\bar{\bar{\mathbf{M}}}\bar{\mathbf{J}}$ is a completely full matrix (computation of all accelerations of motion). Moreover, complementary equations can be found as

$$\mathbf{Q}^T\mathbf{Q}\mathbf{g} = \mathbf{Q}^T(\mathbf{E}-\mathbf{C})^{-1}[\bar{\bar{\mathbf{M}}}(\mathbf{E}-\mathbf{C})^{-1}(\mathbf{J}\ddot{\mathbf{y}}+\xi)+\bar{\mathbf{k}}-\bar{\mathbf{q}}^e]\ , \tag{22}$$

and

$$\mathbf{J}^T\mathbf{J}\ddot{\mathbf{y}} = \mathbf{J}^T(\mathbf{E}-\mathbf{C})\bar{\bar{\mathbf{M}}}^{-1}[(\mathbf{E}-\mathbf{C})^T\mathbf{Q}\mathbf{g}+\bar{\mathbf{q}}^e-\bar{\mathbf{k}}]-\mathbf{J}^T\xi\ . \tag{23}$$

By proper choice of the coordinates, and the frames for the resolution of vectors and matrices, it can be achieved that $\mathbf{Q}^T\mathbf{Q}=\mathbf{E}$ and $\mathbf{J}^T\mathbf{J}=\mathbf{E}$. Thus, the solution of (20) or (21), respectively, is the essential task.

It can be shown, however, that the fxf-inertia matrix **M** is resolvable in triangular form by the Gaussian algorithm and a completely recursive solution can be achieved. Different variations of recursive solutions have been analysed in detail by Valasek [7]. Then, the reactions required are easily available from (22). On the other hand, if all reactions are needed for a more detailed analysis of a rigid-elastic system, (21) can be solved first and the equations of motion obtained from (23).

Simulation of Motion

For the simulation the equation of motion (20) have to be rewritten as a first order differential equation system

$$\begin{bmatrix} \dot{y} \\ \ddot{y} \end{bmatrix} = \begin{bmatrix} \dot{y} \\ \mathbf{M}^{-1}(y)[\ q(y,\dot{y}) - k(y,\dot{y})\ ] \end{bmatrix} . \tag{24}$$

The simulation requires in any case the inversion of the fxf-inertia matrix **M** which can be achieved recursively as mentioned or by using (21) and (23), respectively.

An integration code most successful used in multibody system dynamics is due to Shampine and Gordon [14]. It has been shown by Leister [15] that this predictor-corrector multistep method is more efficient than standard Runge-Kutta codes.

As an example, a three-body pendulum will be treated, featuring most of the nonlinear phenomena in multibody systems to be discovered by simulation.

The pendulum consists of p=3 bodies, Fig. 4. The masses are Mi, the moment of inertia Ii, and GE is the the gravitational acceleration. The pendulum is symmetric with respect to the geometrical parameters: distance O_IC_1 is A, distances O_2C_2, O_3C_3 are B and distances O_IO_2, O_IO_3 are C. The pendulum has f=3 degrees of freedom, the generalized coordinates are introduced as AL_l, l=1(1)3.

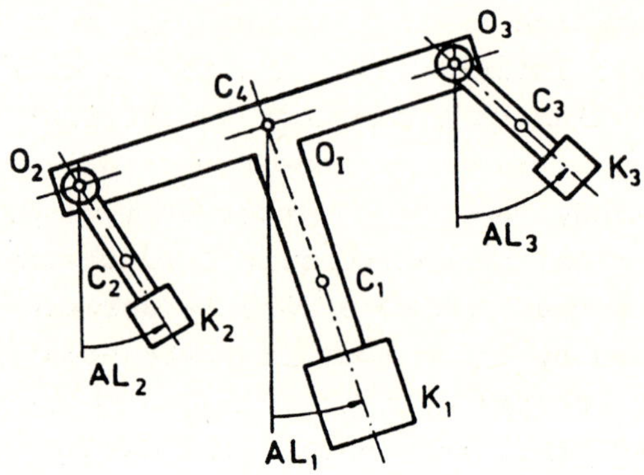

Fig. 4. Three-body pendulum

A broad spectrum of motions can be observed. Each motion is characterized by special initial conditions and system parameters. The equilibrium position is given for vanishing initial conditions $\mathbf{y}_0 = \dot{\mathbf{y}}_0 = \mathbf{0}$. A periodic motion follows with the initial conditions $\mathbf{y}_0 = [0.01 \quad 0.1 \quad 0.1]^T$, $\dot{\mathbf{y}}_0 = \mathbf{0}$.

A chaotic motion is obtained for the initial conditions

$$\mathbf{y}_0 = [3.14 \quad 0 \quad 0]^T, \qquad \dot{\mathbf{y}}_0 = \mathbf{0} \, . \tag{25}$$

The trajectories of the generalized coordinates and the generalized velocities for chaotic motion are shown in Ref. [16] using the parameters M1=2, M2=M3=1,Ii=0, GE=1 and A=2+$\sqrt{3}$, B=C=1.

Animation of Multibody Systems

A convenient verification and visualisation of multibody system simulations is obtained by computer graphics animation. In order to display a broad variety of simulation results for different initial conditions and system parameters, an organisation of the graphic data in a centralized database is necessary.

The concept of geometric modelling and hierarchical data structures is supported by sophisticated graphic software like PHIGS [17]. A hierarchical data organisation allows to group the graphical representation of each system body in a separate structure. The geometric information is defined and stored at the modeling level according to a body-fixed coordinate system.

Since a multibody system consists of a number of discrete bodies, its graphical representation is described by the adequate number of structures. The relationship between the set of structures is defined by the method of instancing. Instancing allows to replicate structures multiple times with different positions, orientations and other attributes. The concept of instancing relates the instances to a homogeneous coordinate representation of transformation matrices, see, e.g. Newman and Sproull [18]. Fig. 5 shows the graphic data organisation of the three-body pendulum and the related structures. Another, more economic procedure would be in this special case of two geometric identical bodies to use only structure 2 and to apply instancing twice.

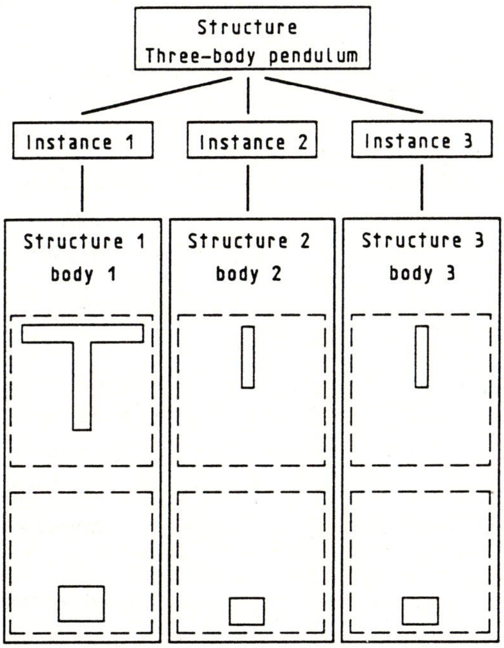

Fig. 5. Graphic data organisation of three-body pendulum

The method of changing a structure's position and orientation by setting an appropiate transformation matrix is an approach to realize a multibody system animation. As a result of the multibody simulation, the 3x1-translation vector and the 3x3-rotation tensor of each body can be obtained numerically for each simulation step. These two quantities are combined to a resulting 4x4 transformation matrix that uniquely describes the new instance of the related graphic object.

This approach is applied for each body and each timestep that has to be displayed. The visualisation of the chaotic motion of the three-body pendulum is shown in Fig.6 for the times indicated.

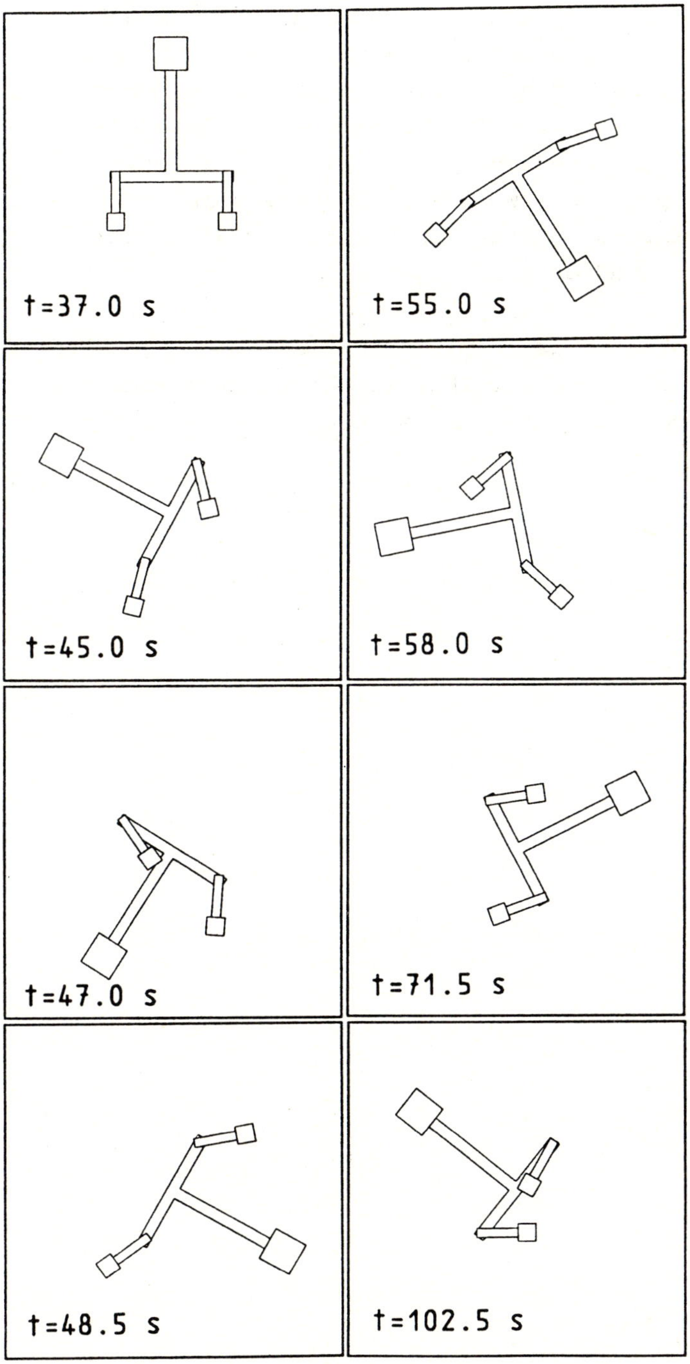

Fig. 6. Animation of chaotic motion

References

1. Magnus, K.(ed.): Dynamics of Multibody Systems. Berlin/...: Springer 1978.
2. Haug, E.(ed.): Computer aided analysis and optimisation of mechanical system dynamics. Berlin/...: Springer 1984.
3. Bianchi, G.; Schiehlen, W.(eds.): Dynamics of Multibody Systems. Berlin/...: Springer 1986.
4. Kreuzer, E.; Schiehlen, W.: Rechnergestütztes Aufstellen der Bewegungsgleichungen gewöhnlicher Mehrkörpersysteme. Ing.-Arch. 46(1977), 185-194.
5. Brandl, H; Johanni, R.; Otter, M.: A very efficient algorithm for the simulation of robots and similar multibody systems without inversion of the mass matrix. In: Theory of Robots. Kopacek, P.; Troch, I.; Desoyer, K.(eds.). Oxford/...: Pergamon Press 1988, 95-100.
6. Führer, C.: Differential-algebraische Gleichungssysteme in Mechanischen Mehrkörpersystemen. Diss. TUM-M8807. München: Technische Universität München 1988.
7. Valasek, M.: On the efficient implementation of multibody system formalisms. IB-17. Stuttgart: Institut B für Mechanik, Universität Stuttgart 1990.
8. Schiehlen, W.(ed.): Multibody Systems Handbook. Berlin/...: Springer 1990.
9. Rauh, J.; Schiehlen, W.: Various approaches for the modeling of flexible robot arms. In: Refined dynamical theories of beams, plates and shells and their applications. Elishahoff, I.; Irretier, H.(eds.). Berlin/...: Springer 1987, 420-429.
10. Rauh, J.: Ein Beitrag zur Modellierung elastischer Balkensysteme. Düsseldorf: VDI-Verlag 1987.
11. Gasch, H.; Knothe, K.: Strukturdynamik, Band 2. Berlin/...: Springer 1989.
12. Schiehlen, W.: Optimal shape of pendulum links. In: Structural optimisation. Rozvany, G.I.N.; Karihaloo, B.L.(eds.). Dordrecht/...: Kluwer 1988, 279-288.
13. Schiehlen, W.: Technische Dynamik. Stuttgart: Teubner 1985.
14. Shampine, L.F.; Gordon, M.U.: Computer solution of ordinary differential equations. San Francisco: Freeman 1975.
15. Leister, G.: Wahl geeigneter Koordinaten zur Dynamikanalyse von Mehrkörpersystemen. ZB-49. Stuttgart: Institut B für Mechanik, Universität Stuttgart 1990.
16. Schiehlen, W.: Modeling and analysis of nonlinear multibody systems. Vehicle System Dynamics 15(1986), 271-288.
17. Brown, M.D.: Understanding PHIGS. San Diego: Template Software Division, 1985.
18. Neman, W.; Sproull, R.: Grundzüge der interaktiven Computergrafik. Hamburg: McGraw-Hill Verlag, 1986.

Contributions to Kinematics, Statics and Dynamics of Multibody Systems Solvable in Linear Computational Complexity

Michael VALÁŠEK, Department of Automatic Control
Vladimir STEJSKAL, Department of Mechanics

Faculty of Mechanical Engineering,
Czech Technical University of Prague
Technicka 4, 166 07 Praha 6, Czechoslovakia

Summary

Computational complexity of solution of all problems of multibody mechanics by traditional methods is $O(n^3)$ where n is number of bodies in the system. Recently there has been developed methods for the solution of direct dynamic problem of multibody systems with tree structure (without loops) which have linear computational complexity $O(n)$. The paper deals with the development of further algorithms for solving other problems of multibody mechanics with the computational complexity possibly only $O(n)$.

1. Introduction

The improvement of computational complexity of solution of dynamical investigation of multibody systems is necessary in order to enable the application of multibody system approach to many more complex practical problems. Recently [4, 5, 2, 3, 1] there have been developed algorithms for the solution of direct dynamic problem of multibody systems without loops with the computational complexity $O(n)$. The algorithm [3] for the solution of direct dynamic problem of multibody systems with loops has the computational complexity $O(n) + O(l^3)$ where l is the number of loops. The development of algorithms which could decrease the computational complexity for multibody systems with loops is very important because such system are often by modelling complex machines. The paper deals with the development [6, 7] of such algorithms not only for dynamics of multibody systems but also for kinematics and statics. All these problems are entirely connected, i.e. the solution of dynamics for a multibody system with loops uses the solution of kinematics for this system.

2. The Geometric Diagram and the Parametric Method in Kinematics

The generalization of old graphical methods [6, 7] in kinematics is based on the combination of the geometric diagram

N. V. Banichuk · D. M. Klimov · W. Schiehlen (Eds.)
Dynamical Problems of Rigid-Elastic
Systems and Structures
IUTAM Symposium Moscow, USSR, May 23-27, 1990

method and the parametric one. The geometric diagram method uses the geometric relation among velocities of different points of multibody system which are constructed in the so called velocity diagram. We fix in space a point O as origin. The velocities of points of interest of investigated multibody system we shall describe as radius vectors with the same origin O. The velocity diagram is a diagram of velocity vectors and their end points. In this way we obtain geometrical interpretation of relations among velocities in the investigated multibody system. The end points of velocities of important points in the investigated multibody system lie in different geometrical sets (e.g. plane straight line) and hence they can be determined by the intersections of these sets. The velocity diagram method is the fundamental method of graphical solution of planar multibody systems [8], but we use its generalization into the spatial multibody systems.

The parametric method uses other facts. In many cases if we knew the velocity of some point of multibody system then we could finish the kinematic solution. Thus we choose the value of this velocity and we finish the kinematic solution. However, such solution is not in general correct because it must fulfill some condition. And from these condition we determine the correct value of the chosen velocity. The chosen values are during the kinematic solution treated as parameters and hence the name of parametric method. The origin of parametric method is also in graphical methods [8].

The geometric diagram and the parametric method are primarly methods of velocity solution of multibody system. There have been derived [6] that the matrix of the system of linear algebraic equations which are to be solved for the solution of position, velocity or acceleration of multibody system are identical. Therefore the geometric diagram and the parametric method can be applied to the solution of position, velocity as well as acceleration of multibody sywstem. Further we shall describe only the solution of multibody system velocity.

The solution of kinematics (e.g. velocities) of a general multibody system is realized according to the following algorithm. A multibody system with dependent variables consists from several kinematic loops. We solve these kinematic loops sequentially. In each loop we can determine 6 dependent variables. However because the loops have mutual bodies and kinematic joints the loops can contain more dependent variables than 6. The values of these dependent variables we must choose and their correct values will be determined from further (e.g. last) kinematic loops which will be solved and where must be fulfilled some conditions in order to be solvable. Each loop is solved by the geometric diagram method.

The open question is the number of the necessary choices for the solution of general multibody system. If the number of these necessary choices is bounded to some fixed finite number this algorithm has the linear computational complexity.

3. The Parametric Method in Statics

The fundamental idea of the parametric method in statics is identical with the parametric method in kinematics. In many cases if we knew the reaction forces of some kinematic joint of multibody system then we could finish the static solution. Thus we choose the value of this force and we finish the static solution. However such solution is not in general correct because it must fulfill some condition. And from these condition we determine the correct value of the chosen force. The chosen values are during the static solution treated as parameters and hence the name of parametric method.

The described algorithm is the most general one for the spatial multibody system. Further promising approach [6] is to construct an equivalent truss to particular multibody system and then to solve only this truss by a special case of the above described algorithm. Because the equivalent truss structurally fulfills many relations among reaction forces of multibody system this approach has promising computational complexity.

The open question is as in kinematics the number of the necessary choices for the solution of general multibody system. If the number of these necessary choices is bounded to some fixed finite number this algorithm has the linear computational complexity.

4. Dynamics

Let us consider a simple kinematic chain with revolute and prismatic joints. Let us use relative coordinates s_i between i-th body and the previous body $p(i)$ in the direction to the frame. We shall describe the kinematic and dynamic quantities by the composite (6,1) vectors and (6,6) matrices [2, 6]. The equations of motion for the i-th body formulated to the hinge point H_i on the axis of the i-th kinematic joint are described by the equation

$$\begin{bmatrix} {}^{i}\mathbf{I}_{i}^{\mathbf{H}} & m_{i}\tilde{\mathbf{d}}_{i} \\ m_{i}\tilde{\mathbf{d}}_{i}^{\mathbf{T}} & m_{i}\mathbf{E} \end{bmatrix} \begin{bmatrix} {}^{i}\alpha_{i} \\ {}^{i}\mathbf{a}_{i}^{\mathbf{H}} \end{bmatrix} = \begin{bmatrix} {}^{i}\mathbf{M}_{i}^{\mathbf{H}} \\ {}^{i}\mathbf{F}_{i}^{\mathbf{H}} \end{bmatrix} - \begin{bmatrix} {}^{i}\omega_{i} \times {}^{i}\mathbf{I}_{i}^{\mathbf{H}}\, {}^{i}\omega_{i} \\ m_{i}\,{}^{i}\omega_{i} \times ({}^{i}\omega_{i} \times \mathbf{d}_{i}) \end{bmatrix} \tag{1}$$

where ${}^{i}\mathbf{I}_{i}^{\mathbf{H}}$ is the traditional tensor of inertia to the hinge point, $\mathbf{d}_{i}$ is the vector from the hinge point to the centre of mass of body i, $\tilde{\mathbf{d}}_{i}$ is its expression as the skewsymetric matrix, m_{i} is the mass of the body i, ${}^{i}\omega_{i}$ is the vector of angular velocity of body i, ${}^{i}\alpha_{i}$ and ${}^{i}\mathbf{a}_{i}^{\mathbf{H}}$ are angular and translational acceleration of body i, ${}^{i}\mathbf{M}_{i}^{\mathbf{H}}$ and ${}^{i}\mathbf{F}_{i}^{\mathbf{H}}$ are the torque and force acting on the body i. All quantities are expressed in the coordinate system firmly attached to the body i with the origin in the hinge point. The equation (1) can be compactly written

$$\mathbf{I}_{i}^{\mathbf{H}}\, \mathbf{a}_{i}^{\mathbf{H}} = \mathbf{f}_{i}^{\mathbf{H}} + \beta_{i}^{\mathbf{H}} \tag{2}$$

where $\beta_{i}^{\mathbf{H}}$ is

$$\beta_{i}^{\mathbf{H}} = - \begin{bmatrix} {}^{i}\omega_{i} \times {}^{i}\mathbf{I}_{i}^{\mathbf{H}}\, {}^{i}\omega_{i} \\ m_{i}\,{}^{i}\omega_{i} \times ({}^{i}\omega_{i} \times \mathbf{d}_{i}) \end{bmatrix} \tag{3}$$

Between the accelerations of bodies of the kinematic chain the recursive relation holds

$$\mathbf{a}_{i}^{\mathbf{H}} = \mathbf{C}_{i}\, \mathbf{a}_{p(i)}^{\mathbf{H}} + \Phi_{i}\, \ddot{s}_{i} + \eta_{i} \tag{4}$$

where

$$\mathbf{C}_{i} = \begin{bmatrix} \mathbf{A}_{i}^{p(i)} & \mathbf{0} \\ -\mathbf{A}_{i}^{p(i)}\tilde{\mathbf{r}}_{i} & \mathbf{A}_{i}^{p(i)} \end{bmatrix} \tag{5}$$

$$\eta_{i} = \begin{bmatrix} (\mathbf{A}_{i}^{p(i)}\, {}^{p(i)}\omega_{p(i)}) \times {}^{i}\omega_{p(i)i} \\ \mathbf{A}_{i}^{p(i)} ({}^{p(i)}\tilde{\omega}_{p(i)}^{2}\, \tilde{\mathbf{r}}_{i} + (\mathbf{A}_{i}^{p(i)}\, {}^{p(i)}\omega_{p(i)}) \times 2\, \dot{s}_{i} \end{bmatrix} \tag{6}$$

and where $\mathbf{A}_{i}^{p(i)}$ is the direction cosine matrix transforming vectors from frame of body $p(i)$ into frame of body i, $\mathbf{r}_{i}$ is the vector from the hinge point $H_{p(i)}$ to the hinge point H_{i} resolved in the frame $p(i)$ and s_{i} is the vector of translation along the axis of the joint i if the kinematic joint i is translational, Φ_{i} is the vector $[0, 0, 1, 0, 0, 0]^{\mathbf{T}}$ if the joint i is rotational and $[0, 0, 0, 0, 0, 1]^T$ if the joint i is translational.

The equations of motion can be derived from Gauss's principle (similar as in [1])

$$\sum_i \partial (\mathbf{a}_i^H)^T (\mathbf{I}_i^H \mathbf{a}_i^H - \beta_i^H) = 0 \tag{7}$$

From (4) we have

$$\partial \mathbf{a}_i^H = C_i \partial \mathbf{a}_{p(i)}^H + \Phi_i \partial \dot{s}_i \tag{8}$$

and substitute into (7)

$$\sum_{i=1}^{n-1} \partial (\mathbf{a}_i^H)^T (\mathbf{I}_i^H \mathbf{a}_i^H - \beta_i^H) + \partial (\mathbf{a}_{n-1}^H)^T C_n^T (\mathbf{I}_n^H \mathbf{a}_n^H - \beta_n^H) +$$
$$+ \partial \dot{s}_n \Phi_n^T (\mathbf{I}_n^H \mathbf{a}_n^H - \beta_n^H) = 0 \tag{9}$$

Because $\partial \dot{s}$ is independent we obtain from (9) the equation of motion

$$\Phi_n^T (\mathbf{I}_n^H \mathbf{a}_n^H - \beta_n^H) = 0 \tag{10}$$

We substitute from (4) into (10) and we express $\dot{s}_n$

$$\dot{s}_n = \mathbf{M}_n^{-1} (\Phi_n^T \beta_n^H - \Phi_n^T \mathbf{I}_n^H (C_n \mathbf{a}_{n-1}^H + \eta_n)) \tag{11}$$

where

$$\mathbf{M}_n = \Phi_n^T \mathbf{I}_n^H \Phi_n \tag{12}$$

Now we substitute (11) into (4), (4) into (9) and modify

$$\sum_{i=1}^{n-2} \partial (\mathbf{a}_i^H)^T (\mathbf{I}_i^H \mathbf{a}_i^H - \beta_i^H) + \partial (\mathbf{a}_{n-1}^H)^T (\mathbf{I}_{n-1}^{H*} \mathbf{a}_{n-1}^H - \beta_{n-1}^{H*}) = 0 \tag{13}$$

where

$$\mathbf{P}_i = E - \mathbf{I}_i^H \Phi_i \mathbf{M}_i^{-1} \Phi_i^T \qquad \gamma_i = \mathbf{P}_i \mathbf{I}_i^H \eta_i - \mathbf{P}_i \beta_i^H \tag{14}$$
$$\mathbf{I}_{p(i)}^{H*} = \mathbf{I}_{p(i)}^H + C_i^T \mathbf{N}_i C_i \qquad \beta_{p(i)}^{H*} = \beta_{p(i)}^H - C_i^T \gamma_i$$

We have obtained the recursive equations of motion for the last body (11) and the transformation of the whole kinematic chain into the kinematic chain without the last body. Hence we can recursively continue until the first body $n = 1$ in the chain. Then the equation (11) is solvable because the acceleration $\mathbf{a}_{p(1)}^H$ is given. Hence we can recursively compute accelerations by (4) and (11) from the first to the last body in the kinematic chain.

Let us apply this way of derivation of equations of motion by the recursive dynamic multibody formalism to the multibody systems with kinematic loops [6]. We must distinguish two kinds of loops:

- external loops which are closed through frame (e.g. Fig. 1a)
- internal loops which are closed by moving bodies (e.g. Fig. 1b)

In the case of external loops (e.g. Fig. 1a) we determine the kinematic solution from which the accelerations of last 5 bodies k are expressed as functions of the acceleration of the

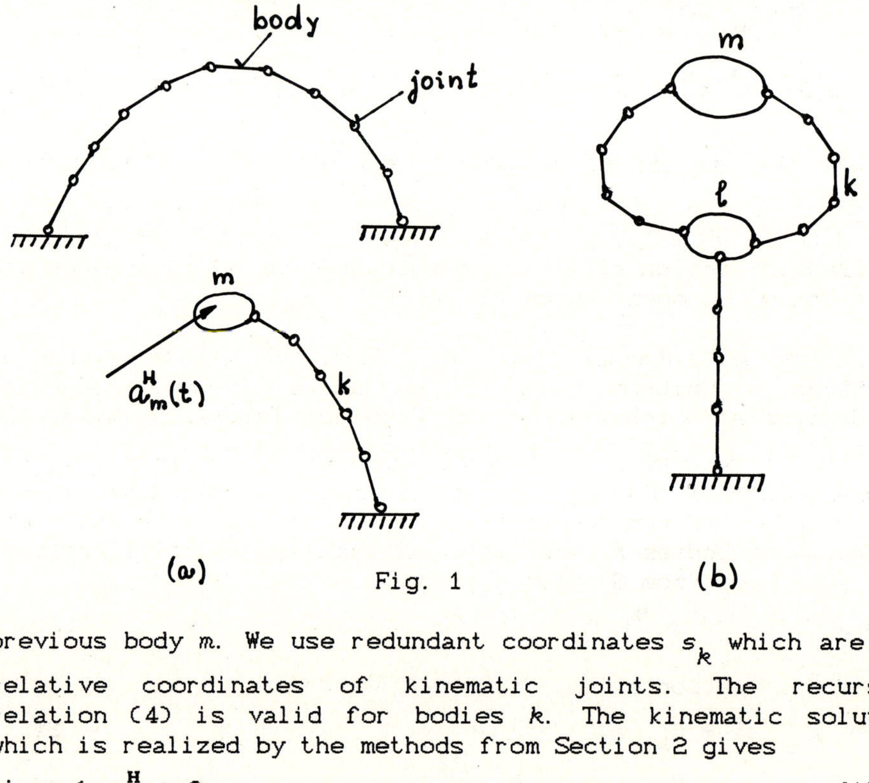

Fig. 1

previous body m. We use redundant coordinates s_k which are the relative coordinates of kinematic joints. The recursive relation (4) is valid for bodies k. The kinematic solution which is realized by the methods from Section 2 gives

$$\dot{s}_k = \mathcal{A}_k \, \mathbf{a}_m^H + \mathcal{R}_k \tag{15}$$

After substitution of (15) into (4) we obtain

$$\mathbf{a}_k^H = C_{km} \, \mathbf{a}_m^H + \eta_{km} \tag{16}$$

$$\partial \, \mathbf{a}_k^H = C_{km} \, \partial \, \mathbf{a}_m^H \tag{17}$$

The equations of motion of such multibody system are also formulated by (7). After modification we have

$$\sum_{i=1}^{m-1} \partial \, (\mathbf{a}_i^H)^T \, (\mathbf{I}_i^H \, \mathbf{a}_i^H - \beta_i^H) + \partial \, (\mathbf{a}_m^H)^T \, (\mathbf{I}_m^H \, \mathbf{a}_m^H - \beta_m^H) +$$

$$+ \sum_k \partial (\mathbf{a}_k^H)^T (\mathbf{I}_k^H \mathbf{a}_k^H - \beta_k^H) = 0 \qquad (18)$$

The difference between (7) and (18) is that the virtual accelerations $\partial \mathbf{a}_k^H$ in (18) are not independent. Therefore we cannot apply the way of derivation of equations of motion directly. We substitute (17) into (18) and modify

$$\sum_{i=1}^{m-1} \partial (\mathbf{a}_i^H)^T (\mathbf{I}_i^H \mathbf{a}_i^H - \beta_i^H) + \partial (\mathbf{a}_m^H)^T ((\mathbf{I}_m^H + \sum_k C_{km}^T \mathbf{I}_k^H C_{km}) \mathbf{a}_m^H -$$

$$+ (\beta_m^H + \sum_k C_{km}^T (-\mathbf{I}_k^H \eta_{km} + \beta_k^H)) = 0$$

$$\sum_{i=1}^{m-1} \partial (\mathbf{a}_i^H)^T (\mathbf{I}_i^H \mathbf{a}_i^H - \beta_i^H) + \partial (\mathbf{a}_m^H)^T (\mathbf{I}_m^{H*} \mathbf{a}_m^H - \beta_m^{H*}) = 0 \qquad (19)$$

Now all the virtual accelerations $\partial \mathbf{a}_i^H$, $\partial \mathbf{a}_m^H$ are independent and thus we can apply the derivation of equations of motion from (7) to (13) in the same way. By this procedure the equations of motion of a loop are transformed to the equations of motion of an open kinematic chain.

In the case of internal loops (e.g. Fig. 1b) the derivation of equations of motion is more difficult. We use redundant coordinates s_k which are the relative coordinates of kinematic joints. We determine the kinematic solution from which the accelerations of 5 bodies k are expressed as functions of the accelerations of the bodies m and l. The recursive relation (4) is valid for bodies k. The kinematic solution which is realized by the methods from Section 2 gives

$$\ddot{s}_k = \mathcal{A}_k \mathbf{a}_m^H + \mathcal{B}_l \mathbf{a}_l^H + \mathcal{R}_k \qquad (20)$$

After substitution of (20) into (4) we obtain

$$\mathbf{a}_k^H = C_{km} \mathbf{a}_m^H + C_{kl} \mathbf{a}_l^H + \eta_{km} \qquad (21)$$

$$\partial \mathbf{a}_k^H = C_{km} \partial \mathbf{a}_m^H + C_{kl} \partial \mathbf{a}_l^H \qquad (22)$$

The equations of motion od such multibody system are also formulated by (7). After modification we have

$$\sum_{i=1}^{l} \partial (\mathbf{a}_i^H)^T (\mathbf{I}_i^H \mathbf{a}_i^H - \beta_i^H) + \sum_{i=l+1}^{m} \partial (\mathbf{a}_i^H)^T (\mathbf{I}_i^H \mathbf{a}_i^H - \beta_i^H) +$$

$$+ \sum_k \partial (\mathbf{a}_k^H)^T (\mathbf{I}_k^H \mathbf{a}_k^H - \beta_k^H) = 0 \qquad (23)$$

The difference between (7) and (23) is that the virtual accelerations $\partial \mathbf{a}_k^H$ in (23) are not independent. Therefore we cannot apply the way of derivation of equations of motion directly. We substitute (21), (22) into (23) and modify

$$\sum_{i=1}^{l} \partial (a_i^H)^T (I_i^H a_i^H - \beta_i^H) + \sum_{i=l+1}^{m} \partial (a_i^H)^T (I_i^H a_i^H - \beta_i^H) +$$

$$+ \sum_k (C_{km} \partial a_m^H + C_{kl} \partial a_l^H)^T (I_k^H (C_{km} a_m^H + C_{kl} a_l^H + \eta_{km}) - \beta_k^H) = 0$$

$$\sum_{i=1}^{l} \partial (a_i^H)^T (I_i^H a_i^H - \beta_i^H) + \sum_{i=l+1}^{m-1} \partial (a_i^H)^T (I_i^H a_i^H - \beta_i^H) +$$

$$+ \partial (a_m^H)^T (I_m^* a_m^H + I_{ml}^H a_l^H - \beta_m^{H*}) +$$

$$+ \partial (a_l^H)^T (I_{lm}^H a_m^H + I_l^{H*} a_l^H - \beta_l^{H*}) = 0 \qquad (24)$$

where

$$\beta_m^{H*} = \beta_m^H + \sum_k C_{km}^T \beta_k^H - \sum_k C_{km}^T I_k^H \eta_{km}$$ and similarly for β_l^{H*}.

$$I_m^* = I_m + \sum_k C_{km}^T I_k^H C_{km} \qquad (25)$$

$$I_{ml}^H = \sum_k C_{km}^T I_k^H C_{kl}$$ and similarly for I_{lm}^H.

$$I_l^{H*} = \sum_k C_{kl}^T I_k^H C_{kl}$$

The relations (4) and (8) are also valid for body m. We substitute these relations into (24)

$$\sum_{i=1}^{l} \partial (a_i^H)^T (I_i^H a_i^H - \beta_i^H) + \sum_{i=l+1}^{m-1} \partial (a_i^H)^T (I_i^H a_i^H - \beta_i^H) + \qquad (26)$$

$$+ (C_m \partial a_{p(m)}^H + \Phi_m \partial \dot{s}_m)^T (I_m^* (C_m a_{p(m)}^H + \Phi_m \dot{s}_m + \eta_m) +$$

$$+ I_{ml}^H a_l^H - \beta_m^{H*}) + \partial (a_l^H)^T (I_{lm}^H a_m^H + I_l^{H*} a_l^H - \beta_l^{H*}) = 0$$

Because $\partial \dot{s}_m$ is independent we obtain from (26) the equation of motion

$$\Phi_m^T (I_m^{H*} (C_m a_{p(m)}^H + \Phi_m \dot{s}_m + \eta_m) + I_{ml}^H a_l^H - \beta_m^{H*}) = 0 \qquad (27)$$

We designate

$$G_m = \Phi_m M_m^{-1} \Phi_m^T \qquad (28)$$

Then from (27)

$$\dot{s}_m = - M_m^{-1} \Phi_m^T I_m^* C_m a_{p(m)}^H + M_m^{-1} \Phi_m^T (\beta_m^{H*} - I_m^* \eta_m) - M_m^{-1} \Phi_m^T I_{ml}^H a_l^H \qquad (29)$$

and from (4)

$$a^H_m = P^T_m C_m a^H_{p(m)} + P^T_m \eta_m + G_m (\beta^{H*}_m - I_{ml} a^H_l) \qquad (30)$$

These expressions are substituted back to the equations of motion (26) and after modification we obtain

$$\sum_{i=1}^{l} \partial (a^H_i)^T (I^H_i a^H_i - \beta^H_i) + \sum_{i=l+1}^{m-2} \partial (a^H_i)^T (I^H_i a^H_i - \beta^H_i) +$$
$$+ \partial (a^H_{p(m)})^T (I^{H*}_{p(m)} a^H_{p(m)} + I_{p(m)l} a^H_l - \beta^{H*}_{p(m)}) + \qquad (31)$$
$$+ \partial (a^H_l)^T (I_{lp(m)} a^H_{p(m)} + I^{H*}_l a^H_l - \beta^*_l) = 0$$

where

$$I^{H*}_{p(m)} = I^H_{p(m)} + C^T_m I^*_m P_m C_m \qquad I^{H*}_l = I^{H*}_l - I^H_{lm} G_m I^H_{ml}$$

$$I_{p(m)l} = C^T_m P_m I^H_{ml} \qquad \text{and similarly } I_{lp(m)}.$$

$$\beta^{H*}_{p(m)} = \beta^H_{p(m)} - C^T_m (I^*_m P^T_m \eta_m - P_m \beta^{H*}_m) \qquad (32)$$
$$\beta^{H*}_l = \beta^{H*}_l - I^H_{lm} P^T_m \eta_m - I^H_{lm} G_m \beta^*_m$$

The equation of motion (30) has the same structure as the equation of motion (24) however recursively reduced by one body m. his recursion can continue until $p(m) = l$. Then instead of equation (30) we obtain

$$\sum_{i=1}^{l-1} \partial (a^H_i)^T (I^H_i a^H_i - \beta^H_i) + \partial (a^H_l)^T (I^{H*}_l a^H_l - \beta^*_l) = 0 \qquad (33)$$

This equation of motion is exactly the same as the equation of motion (7) for the pure kinematic chain and the further derivation is the same as above.

If the computational complexity of kinematic solution is linear and the number of accelerations on which the dependent acceleration a^H_k depends (e.g. in equation (21) two) is bounded by some fixed finite number then the computational complexity of dynamic solution is also linear.

5. Conclusion

The described algorithms decrease the computational complexity of the solution of kinematic, static and dynamic problems of multibody system rapidly. However the proof of their linear computational complexity for general multibody system is still an open question. Further investigation are also necessary in order to study the application of these algorithms to singular cases in mechanics of multibody systems.

Acknowledgements

The results of this study have been obtained during the research stay of Michael Valášek at the Institute B für Mechanik, Universität Stuttgart , FRG supported by the Alexander von Humboldt Foundation. The author wishes to acknowledge this support.

References

1. Bae, D.S.; Haug, E.J.: A Recursive Formulation for Constrained Mechanical System Dynamics: Part I: Open-Loop Systems, Part II: Closed Loop Systems, Mech. Structures and Machines, 15(1987), (3,4), 359-382, 481-506

2. Brandl, H.; Johanni, R.; Otter, M.: A Very Efficient Algorithm for the Simulation of Robots and Similar Multibody Systems Without Inversion of the Mass Matrix, In: Kopacek, P.; Troch, I.; Desoyer, K. (eds.): Theory of Robots, Proc. of the IFAC/IFIP/IMACS International Symposium, Vienna 1986, Pergamon Press, Oxford 1988

3. Brandl, H.; Johanni, R.; Otter, M.: An Algorithm for the Simulation of Multibody Systems With Kinematic Loops, In: Bautista, E.; Garcia-Lomas, J.; Navarro, A. (eds.): The Theory of Machines and Mechanisms, Proc. of the 7th IFToMM World Congress, Pergamon Press, Oxford 1987

4. Featherstone, R.: The Calculation of Robot Dynamics Using Articulated Body Inertias, The International Journal of Robotics Research, 2, No. 1, 13-30

5. Featherstone, R.: Robot Dynamics Algorithms, Kluwer Academic Publishers, Dodrecht 1987

6. Valášek, M.: Contributions to Kinematics, Statics and Dynamics of Multibody Systems Solvable in Linear Computational Complexity, IB-18,ISBN 3-927618-06-3, Institut B für Mechanik, Universität Stuttgart, Stuttgart 1990

7. Valášek, M.: Differential Kinematic Analysis of General 6R Robot Manipulators Without the Inverse of Jacobian Matrix, Advances in Robot Kinematics, Linz 1990 (to appear)

8. Wittenbauer, F.: Graphische Dynamik, Springer Verlag, Berlin 1923